General Physics 202/204
Laboratory Manual

Pennsylvania State University
Physics Department

Edited by
R.M. Herman
with the assistance of
A. Titus

KENDALL/HUNT PUBLISHING COMPANY
4050 Westmark Drive Dubuque, Iowa 52002

Copyright © 1993 by Pennsylvania State University

ISBN 0-8403-8347-9

All rights reserved. No part of this publication may be reproduced,
stored in a retrieval system, or transmitted, in any form or by any
means, electronic, mechanical, photocopying, recording, or otherwise,
without the prior written permission of the copyright owner.

Printed in the United States of America
10 9 8 7 6 5 4

TABLE OF CONTENTS

Acknowledgements..i
Elementary Physics Laboratories - Introduction.....................ii
Experimental Errors...iv
Graphical Presentation and Interpretation of Data................xx

Physics 202 Experiments:

1. Motion of a Freely Falling Body 1
2. Experimental Error - Pendulum 5
3. Experimental Errors - VTVM 8
4. Electric Field Plotting 11
5. D. C. Circuits, Part I: Components 15
6. D. C. Circuits, Part II: EMF, Internal Resistance and D. C. Power ... 20
7. Resistance Measurements - Slide Wire Wheatstone Bridge 23
8. Temperature Coefficient of Resistance 26
9. The Cathode Ray Oscilloscope 32
10. Magnetic Field Due to a Circular Coil 38
11. Relaxation Oscillator 41
12. Current Balance 45
13. Elements of A.C. Circuits 49
14. A. C. Series Circuits - Resonance and Forced Oscillation 54
15. Computerized Pendulum 59

Physics 204 Experiments:

1. Transverse Waves 62
2. Longitudinal Waves 67
3. Dispersion of Glass 73
4. The Michelson Interferometer 82
5. Interference - Newtons Rings 90
6. Diffraction of Light 98
7. Diffraction and the Wavelength of Light 106
8. The Diffraction Grating Spectrometer 109
9. Polarization of Light 114
10. Holography 124
11. Charge to Mass Ratio of the Electron 132
12. Millikan's Oil Drop Experiment 137
13. The Franck- Hertz Experiment 142
14. Nuclear Physics 147

Acknowledgements

The efforts of T. Accordino, A. Danko and S. Roberts in preparing this text were indispensable. Members of the Penn State Society of Physics Students helped in revisions of some of the experiments.

ELEMENTARY PHYSICS LABORATORIES

INTRODUCTION

The experiments that you will study in the laboratory have been selected to illustrate certain basic principles and methods of physics. In particular, these experiments have certain objectives which include:

1. Acquainting you with some of the techniques and basic apparatus used in experimental work.

2. Demonstrating in the laboratory certain physical concepts and principles.

3. Introducing you to the methods of data analysis.

4. Giving you a feeling for the relative worth of your experimental results, and introducing you to the methods of error analysis.

Because there are a large number of students enrolled in the physics laboratory, a certain amount of regimentation is necessary in order for the laboratory to run smoothly.

A. Laboratory Attendance

You will be expected to attend the laboratory at your officially scheduled time. Make-up labs are not permitted except in unusual circumstances; so be careful not to oversleep or "forget"! If you have a legitimate reason for not attending at the scheduled time, you should arrange for a make-up as follows. Give your instructor a note stating the lab to be missed and the reason for missing it. Then you and the instructor will determine if and when you might make up the lab and the instructor will tell you what to do next. Except in the case of a sudden emergency (car crash, etc.) you are expected to request a make-up **before** you miss the lab.

In any event, missed labs must be made up within one week after the date of the missed lab period. You will not be accepted into any lab other than the one you are scheduled for unless the proper arrangements have been made by your instructor. The above regulations are necessary to prevent overcrowding of the labs by students making up missed experiments.

B. Preparation for Laboratory Experiments

The experiments you will do in the laboratory are designed to be completed in the laboratory period with the lab report turned in at the end of the period by students who have adequately prepared for the experiments. It is expected that you will have thoroughly read the write-up for the experiment. The properly prepared student will arrive at laboratory with the following things prepared ahead of time. You will have a general idea of the purpose of the experiment and the operation of the equipment to be used. You will know which equations you will be using to calculate your results and you will know how to calculate from the data the size of the errors to be assigned to your results. Your laboratory instructor may ask to see evidence of the preparation mentioned above before you start the lab and grade them for use as a quiz grade. He also may give quizzes on the material which you are to have prepared.

EXPERIMENTAL ERRORS

INTRODUCTION AND MOTIVATION

"In this world, nothing is certain but death and taxes." -- Benjamin Franklin.

"Absolute certainty is a privilege of uneducated minds - and fanatics. It is, for scientific folk, an unattainable ideal." -- Cassius Kuyser.

No measurement made in the laboratory is ever perfectly exact. Therefore, one always wishes to know not only the measured magnitude of a physical quantity determined in the laboratory, but also with what accuracy the quantity has been determined.

For example, if someone has measured the diameter of a wire as 0.0323 cm, we want to know whether the measurement is good to 1 part in 1000 (0.1%), 1 part in 100 (1%), or 1 part in 10 (10%), etc. Whatever the error may be, we want to know, at least approximately, what its size is. Knowing the uncertainties in ones data will allow decisions about future actions to be made more successfully.

For this reason, the following discussion of uncertainties in experimental measurements is provided. This is a subject of great importance. The methods and terminology described below may seem new and strange to you, but they are used daily in all practical scientific research and engineering work. Thus it is to your advantage to learn these methods. You will have frequent opportunities to practice applying these methods to a variety of laboratory situations.

One method of error analysis (not the one we will use) is based upon the principles of mathematical statistics. Unfortunately, statistical methods can only be meaningfully applied when one has large amounts of data for a given system. In many cases, especially introductory physics labs, these large quantities of data are not available; the available data are often limited to a single set of readings, or at most a few sets; then statistical methods are not applicable, and some other methods must be devised.[1] All of what follows will be devoted to methods of answering the following question: Given a single set of data, how can its accuracy be estimated?

DEFINITION OF ERROR

Physical measurements involve using a set of apparatus and operations in an attempt to assign a numerical value to the quantity being measured. (Example: Here is a metal wire. What number shall we use to represent its diameter in centimeters?) There is always an element of uncertainty about the numerical value. The error (ε) in the measured value is

---

[1] If you are curious about the statistical method of treatment of errors, see the discussion and references in Chapter I of: Thomas Brown (Ed.), The Lloyd William Taylor Manual of Advanced Undergraduate Experiments in Physics, Addison-Wesley, Reading MA.

simply defined as the difference between the value resulting from measurement (Z_{MEAS}) and the 'true' value of the quantity (Z_{TRUE}).

$$\varepsilon = Z_{MEAS} - Z_{TRUE} \qquad (1)$$

In this context, the term "error" carries no implication of a mistake or blunder. Note that ε can be either positive or negative, depending on whether Z_{MEAS} is larger or smaller than Z_{TRUE}.

It usually happens that we can estimate the approximate magnitude of the error, $|\varepsilon|$, but cannot determine the sign of the error. With this limited information, the best we can hope to do is "bracket" the value of Z_{TRUE} between two numbers Z_{MIN} and Z_{MAX} such that

$$Z_{MIN} < Z_{TRUE} < Z_{MAX} \; . \qquad (2)$$

The result of an experimental measurement is usually quoted in the form

$$Z_{TRUE} = Z_{MEAS} \pm \Delta Z \qquad (3)$$

where ΔZ represents a positive number. ΔZ is chosen in such a way that there is a high probability that

$$|\varepsilon| \leq \Delta Z \qquad (4)$$

i.e., that

$$|Z_{TRUE} - Z_{TRUE}| \leq \Delta Z \; . \qquad (5)$$

Note the use of the word "probability". We can never be completely sure that the values we obtain for Z_{MEAS} and ΔZ really satisfy eq. (5). We can only be more or less sure.

In the remainder of this essay (and also in technical literature in general), the term "error" is frequently used for ΔZ. But in a given situation it is usually clear from the context whether the term "error" is referring to ε or ΔZ. A more appropriate term for ΔZ might be the "uncertainty", since ΔZ is a measure of how uncertain we are about the true value of Z.

DISCREPANCY

When a quantity has been measured by two different individuals, the difference between the two values is called the <u>discrepancy</u>.

For example, the handbook value for the temperature coefficient of resistance for copper is $\alpha_{20} = 3.9 \times 10^{-3}$ °C^{-1}. In the Wheatstone bridge experiment, you might obtain a value of $\alpha_{20} = 3.7 \times 10^{-3}$ °C^{-1}. The discrepancy between your value and the handbook value is 0.2×10^{-3} °C^{-1} (about 5%).

QUOTING ERRORS: ABSOLUTE ERROR AND RELATIVE ERROR

We use the value of ΔZ to help us decide "how good" our value for Z is. For some purposes, we are interested in the <u>absolute error</u> (ΔZ) and quote our answer as $Z_{MEAS} \pm \Delta Z$. (Example: L = 73.02 ± 0.05 cm).

Another way of quoting the error (and one that is often more revealing) is to give the <u>relative error</u> ($\Delta Z/Z$). The relative error tells us how large the error is in comparison with the measured quantity.

For instance, we might measure the thickness (T) of a 25 cent piece with a millimeter scale and determine that T = 1.3 ± 0.4 mm. The error ($\Delta T = 0.4$ mm) is small on the human scale, but is large when compared with the measured quantity (the thickness of the coin):

$$\frac{\Delta T}{T} = \frac{0.4 \text{ mm}}{1.3 \text{ mm}} \approx 0.3 \approx 30\% \quad !$$

To emphasize the relative error you might write T = 1.3 mm ± 30%.

For the results obtained in elementary physics experiments, the relative error is often a few <u>parts per hundred</u> (i.e., a few <u>percent</u>). In other words, ΔZ is a few hundredths of Z. More accurate measurements are often quoted not in percent but in <u>parts per thousand</u>, <u>parts in 10^4</u>, <u>parts per million (ppm)</u>, etc. A manufacturer may guarantee that the value he quotes for the resistance (R) of a precision resistor is "accurate to 20 parts per million". This is shorthand for $\Delta R/R = 20 \times 10^{-6}$.

The absolute error is easily calculated from the relative error using the formula

$$\Delta Z = \left(\frac{\Delta Z}{Z}\right) Z \qquad (6)$$

If a carbon resistor is rated at R = 400 Ohm ± 5%, then $\Delta R/R = 5 \times 10^{-2}$, while $\Delta R = 5 \times 10^{-2}$ R = 20 Ohm. Hence,

R = 400 ± 20 ohm.

ACCURACY AND PRECISION

The terms "accuracy" and "precision" have sharply different meanings. A measurement of the quantity Z is said to be highly <u>accurate</u> if Z_{MEAS} is very close to Z_{true}. On the other hand, a measurement of Z is said to be highly <u>precise</u> if repeated determinations of Z all give very nearly the same value of Z_{MEAS}. High precision is usually a prerequisite for high accuracy; but it is possible to have a highly precise measurement that is not very accurate at all.

SOURCES OF ERROR: SYSTEMATIC ERRORS AND INDETERMINATE ERRORS

Errors in experimental measurements arise from a multitude of sources. Usually errors are divided into two classes on the basis of whether or not the sign of the error can (in principle) be determined. If the sign of the error can be determined (at least in principle), then the error is said to be <u>systematic</u> or <u>determinate</u>. For purposes of discussion, these systematic errors may be loosely divided into three groups: instrumental, personal, and external. Examples of sources of <u>systematic instrumental error</u> are: a watch that always "loses" four seconds per hour; a micrometer that reads 0.02 mm (instead of zero) when the jaws are together; an improperly calibrated meter stick; a supposedly pure sample that actually contains impurities; a meter stick that gives all measurements as 2% shorter than the true length (calibration error). <u>Systematic personal errors</u> are due to some definite bias of the observer. For example, the observer may always estimate too high when estimating readings between scale divisions, or may always tighten the micrometer jaws too hard when making measurements with a micrometer. <u>External systematic errors</u> result from systematic changes in external conditions (wind, temperature, humidity, etc.). Examples are the expansion of a metal scale as the temperature increases, and the swelling of a meter stick as the humidity increases.

If it is not possible (even in principle) to determine the sign of the error, then the error is said to be <u>indeterminate or random</u>. For example, when estimating the position of a point between marked divisions on a millimeter scale, it is usually assumed that the estimate is just as likely to be too high as too low; i.e., the scale reading error is usually assumed to be indeterminate, as are all other errors caused by non-constancy of the senses of the observer, or random effects which may influence the measurement.

The ideas of systematic and indeterminate error have the following significance: the effects of indeterminate errors can be reduced somewhat by measuring the same quantity many times and taking an average. Sometimes the error will be positive, sometimes negative and, hopefully, when the data is averaged the negative errors will cancel the positive errors to some extent, and the average will hopefully be more accurate than an individual measurement. On the other hand, averaging will not have the same effect on systematic errors. The systematic errors from a given source always have the same sign and thus will not cancel when the average is taken.

ERRORS IN DIRECT MEASUREMENTS

By a <u>direct measurement</u> we mean a measurement that is made with the use of a single piece of apparatus, as opposed to a result that is calculated on the basis of measurements made with several different pieces of apparatus.

For example, a direct measurement of the volume of a solid cylinder can be obtained by immersing the cylinder in a liquid and determining how much liquid is displaced. An <u>indirect</u> method of determining the volume is to measure the length and diameter of the cylinder and then calculate the volume using the formula $V = \pi D^2 L/4$.

Even a direct measurement is usually not completely "direct". Usually a measurement of the desired quantity (X) is obtained by taking two readings (R_1 and R_2) on a scale and then determining the difference of the readings:

$$X = R_2 - R_1 \ . \tag{7}$$

We want to know how to estimate the error (ΔX) associated with the value of X thus obtained. Now, each reading is uncertain by some amount (ΔR_1 and ΔR_2) and there is also a calibration error to consider. Here we assume that the calibration error is indeterminate and hence cannot be corrected. The calibration error is proportional to the <u>difference in scale readings</u>, that is $\Delta C |R_1 - R_2|$. These errors are a type of systematic instrumental error. It is easy to determine a general formula for ΔX by considering the extreme values of X that would be obtained from the values of the readings $R_1 \pm \Delta R_1$, $R_2 \pm \Delta R_2$ and adding in the calibration error. For X given above,

$$[\text{Highest value of X}] = R_2^{MAX} - R_1^{MIN} + \Delta C |R_2 - R_1|$$

$$= (R_2 + \Delta R_2) - (R_1 - \Delta R_1) + \Delta C |R_2 - R_1|$$

or

$$[\text{Highest value of X}] = X + (\Delta R_2 + \Delta R_1 + \Delta C |R_2 - R_1|) \ .$$

In the same way we obtain

$$[\text{Lowest value of X}] = X - (\Delta R_2 + \Delta R_1 + \Delta C |R_2 - R_1|) \ ,$$

so

$$\Delta X = \Delta R_2 + \Delta R_1 + \Delta C |R_2 - R_1| \ , \tag{8}$$

with relative error

$$\frac{\Delta X}{X} = \frac{\Delta R_2 + \Delta R_1}{|R_2 - R_1|} + \Delta C \ . \tag{9}$$

Notice that the errors are added together, not subtracted.

The following example will show how these formulas are applied in practice and will also point out practical methods of minimizing the errors in direct measurements. This example deals with a specific instrument; however, the conclusions drawn are applicable to more than just a single instrument. You should note how information obtained from this experience can be used in other situations.

Consider the situation in the "falling body" experiment, where you want to measure the distance between dots on a waxed paper tape. Suppose that you place the 2 cm marker of the scale directly upon one of the dots and also read where the other dot lies on the scale. You are using a scale with markings 1 mm (0.1cm) apart. Due to fuzziness of the dots and the problem of estimation between scale divisions, you might decide that the position of a dot can only be located on the scale to within ±0.05cm.

If your readings are $R_1 = 2.00 \pm 0.05$ cm and $R_2 = 43.26 \pm 0.05$ cm and if the scale calibration is accurate to $\pm 0.05\%$, (i.e., $\pm 0.05\%$ of the length measured, that is, $\Delta C = 5 \times 10^{-4}$), then according to eqs. (7) and (8)

$$S = R_2 - R_1 = 43.26 \text{ cm} - 2.00 \text{ cm} = 41.26 \text{ cm}$$

$$\Delta S = \Delta R_2 + \Delta R_1 + \Delta C \, S$$

$$= 0.05 \text{ cm} + 0.05 \text{ cm} + 5 \times 10^{-4} \times 41 \text{ cm}$$

$$\approx 0.12 \text{ cm}$$

while

$$\frac{\Delta S}{S} = \frac{0.12 \text{cm}}{41 \text{cm}} \approx 0.003 = 0.3\% \quad .$$

You would quote your measurement as

$$S = 41.26 \pm 0.12 \text{ cm or } S = 41.26 \text{ cm} \pm 0.3\%.$$

Note the following:

1. The input data for ΔR_1, ΔR_1, and ΔC are usually only good to one significant figure. In such cases, you only need to use one or two significant figures in your calculation and statement of the error. (For example, 41.26 cm is rounded off to 41 cm when calculating ΔS and $\Delta S/S$. Furthermore, the error is quoted as 3%, not 2.93%). The idea of

"significant figure" is discussed more fully in White, M. W. and Manning, K. V., <u>Experimental College Physics</u>, 3rd ed., McGraw-Hill, New York, 1954, pp. 12 ff.

2. The calibration error in this example (0.02 cm) is much smaller than the total reading error (0.10 cm) – five times smaller, in fact. Since the errors in this case are simply added together (not multiplied), we could have dropped this small term and still have come out with about the same result for ΔS and $\Delta S/S$. This dropping of negligible contributions makes the computation simpler and is often done. (Of course, before you ignore a term, you must be certain that it is indeed small compared to the other terms in the sum.)

3. By noting how much each source of error contributes to ΔS, you can see where you should start if you want to make your measuring technique more accurate. For example, in the case just considered the reading error (0.10 cm) is much larger than the calibration error (0.02 cm), so to reduce ΔS, your first step would be to look for a way to reduce the reading error. For example, you might try to make the marks less fuzzy by changing your experimental technique, or you might try a scale with more closely spaced divisions, or a vernier caliper.

4. Suppose the same scale is used to measure various distances. If the reading errors are always the same (e.g., ± 0.05 cm) then smaller values of S will have larger relative errors associated with them.

For the situation in the example above,

$$\frac{\Delta S}{S} \approx \frac{0.1 \text{cm}}{S}$$

so as the size of S decreases towards 0.1 cm, $\Delta S/S$ becomes very large. For example, the error is measuring the distance between two points that are about 1 cm apart would be

$$\frac{\Delta S}{S} \approx \frac{0.1 \text{cm}}{1 \text{cm}} = 0.1 = 10\% \quad .$$

This is a large relative error, even though the absolute error (~0.1 cm) is small on the human scale.

PROPAGATION OF INDETERMINATE ERRORS

A. Introduction

Often we cannot <u>directly</u> measure the physical quantity in which we are interested. Fortunately, the quantity of interest is usually related (by some known equation) to one or more quantities that are measurable. The procedure is to determine the measurable quantities and then calculate the desired quantity using the known equation. For instance, the resistance (R) of a current-carrying carbon resistor can be determined by measuring the current (I) through the resistor and the voltage (V) across the resistor and then calculating R from the well-known equation $R = V/I$. Now, each of the directly measured quantities used in the calculation will be uncertain by some amount; these uncertainties will cause the calculated result to also be uncertain by some amount. In the remainder of this section, you will learn how to determine the error in the calculated quantity when the error in each of the measured quantities is known.

Let $F = F(X,Y)$ represent the quantity calculated from the measured quantities X and Y. (Example: $R(I,V) = V/I$) Let ΔF represent the error in F that results from the errors ΔX and ΔY. Here, for the sake of clarity, only two measured quantities (X and Y) are considered. The method of extension to more variables is straightforward and should be obvious.

B. First Method

One obvious way to calculate F is to consider the extreme values of F that would result from using extreme values of X and Y. Suppose you want to determine the resistance of a carbon resistor using $R = V/I$. If the measured values and their errors are $V = 1.45 \pm 0.05$ volt and $I = 0.225 \pm 0.004$ amp, then your calculated value for the resistance would be

$$R = \frac{V}{I} = \frac{1.45 \text{ volt}}{0.225 \text{ amp}} = 6.44 \text{ ohm} \quad \text{Note:} \left(1 \text{ ohm} \equiv 1 \frac{\text{volt}}{\text{amp}}\right)$$

The maximum value of R (R_{MAX}) consistent with ΔV and ΔI will occur when the numerator is largest and the denominator is smallest,

$$R_{MAX} = \frac{V_{MAX}}{I_{MIN}} = \frac{V + \Delta V}{I - \Delta I} = \frac{1.45 + 0.05 \text{ volt}}{0.225 - 0.004 \text{ amp}} = \frac{1.50 \text{ volt}}{0.221 \text{ amp}} = 6.79 \text{ ohm}.$$

Similarly, to obtain R_{min} the opposite condition is used.

$$R_{MIN} = \frac{V - \Delta V}{I + \Delta I} = \frac{1.45 - 0.05 \text{ volt}}{0.225 + 0.004 \text{ amp}} = \frac{1.40 \text{ volt}}{0.229 \text{ amp}} = 6.12 \text{ ohm}$$

Now

$$\Delta R^+ \equiv R_{MAX} - R = 6.79 \text{ ohm} - 6.44 \text{ ohm} = 0.35 \text{ ohm}$$

$\Delta R^- \equiv R - R_{MIN} = 6.44 \text{ ohm} - 6.12 \text{ ohm} = 0.32 \text{ ohm}.$

Therefore,

$$\Delta R \approx 0.3 \text{ ohm}, \text{ or } R = 6.44 \pm 0.3 \text{ ohm}$$

and

$$\frac{\Delta R}{R} \approx \frac{0.3 \text{ ohm}}{6.12 \text{ ohm}} \approx 0.05 = 5\%, \text{ or } R = 6.44 \text{ ohm} \pm 5\%.$$

This method is completely legitimate, but it obviously involves some computation, even for this simple example. Fortunately, there is a simpler method available.

C. Second Method

1. <u>Derivation</u>. Actually, the second method is based on the same basic principle as the first method: one determines the extreme values of F consistent with $\pm\Delta X$ and $\pm\Delta Y$; then ΔF is the difference between F and the extreme values. It will soon become clear that the main advantage of this second method described below is that it allows us to avoid actually calculating the extreme values numerically, thus saving us a lot of work.

The value of ΔF is produced by $\pm\Delta X$ and $\pm\Delta Y$ and can be obtained in a straightforward way by writing a Taylor expansion of F(X,Y) about the point X_0, Y_0, where X_0 and Y_0 represent the measured values of X and Y.

$F \pm \Delta F \equiv$ Extreme values of $F(X_0 \pm \Delta X, Y_0 \pm \Delta Y)$

$$= \left\{ F(X_0, Y_0) + \left(\frac{dF}{dX}\right)_{\text{fixed Y}} \cdot (\pm \Delta X) + \left(\frac{dF}{dY}\right)_{\text{fixed X}} \cdot (\pm \Delta Y) + \begin{pmatrix} \text{second-order} \\ \text{and higher-order} \\ \text{terms} \end{pmatrix} \right\}_{\substack{\text{extreme} \\ \text{values}}}$$

(10)

By comparing terms, we see that

$$F = F(X_0, Y_0)$$

and, neglecting the second and higher order terms,

$$\Delta F = \left\{ \left(\frac{dF}{dX}\right)_{\text{fixed Y}} \cdot (\pm \Delta X) + \left(\frac{dF}{dY}\right)_{\text{fixed X}} \cdot (\pm \Delta Y) \right\}_{\substack{\text{largest} \\ \text{value}}} \quad (11)$$

In eq. (11), the largest value of the total expression enclosed in curly braces occurs when each term in the sum is positive. So when F is a function that has (dF/dX) positive, we would use $+\Delta X$. On the other hand if F is a function that has ((dF/dX) negative, we would use $-\Delta X$, so that the product would be positive. We can represent this by inserting absolute value signs, in which case the $\pm$ signs become superfluous and can be dropped.

$$\Delta F = \left| \left(\frac{dF}{dX}\right)_{\text{fixed Y}} \cdot (\Delta X) \right| + \left| \left(\frac{dF}{dY}\right)_{\text{fixed X}} \cdot (\Delta Y) \right| \quad (12)$$

To illustrate this, consider the previous example, where R(V,I)=V/I, with V = 1.45 ± 0.05 volt and I = 0.225 ± 0.004 amp. The answers obtained by the method used there were R = 6.44 ohm, $\frac{\Delta R}{R} \approx 5\%$, $\Delta R \approx 0.3$ ohm. For the sake of comparison, we now repeat the calculation of R and of $\Delta R/R$ and ΔR, this time using the result just obtained, eq. (12). The calculation of R itself goes exactly as before,

$$R = R(I_0, V_0) = \frac{V_0}{I_0} = \frac{1.45 \text{ volt}}{0.225 \text{ amp}} = 6.44 \text{ ohm}$$

with eq. (12) then giving

$$\Delta R = \left| \left(\frac{dR}{dV}\right) \cdot (\Delta V) \right| + \left| \left(\frac{dR}{dI}\right) \cdot (\Delta I) \right|$$

$$= \left| \left(\frac{1}{I}\right) \cdot (\Delta V) \right| + \left| \left(\frac{-V}{I^2}\right) \cdot (\Delta I) \right| +$$

or

$$\Delta R = + \frac{1}{I_0} \cdot \Delta V + \frac{V_0}{I_0^2} \cdot \Delta I \quad (13)$$

Note that the minus sign in $-V/I^2$ disappears when taking the absolute value. This result for ΔR is satisfactory as it stands, but numerical calculation can be simplified by using the following trick (useful only when F is a product or a quotient), namely factoring R=V/I itself from the above results, to obtain

$$\Delta R = \frac{V}{I}\left(\frac{\Delta V}{V} + \frac{\Delta I}{I}\right) \tag{14}$$

or

$$\frac{\Delta R}{R} = \frac{\Delta V}{V} + \frac{\Delta I}{I} \tag{15}$$

Then, using the numbers given above

$$\frac{\Delta R}{R} = \frac{0.004 \text{ amp}}{0.225 \text{ amp}} + \frac{0.05 \text{ volt}}{1.45 \text{ volt}} = 0.018 + 0.035 \cong 0.05 = 5\%$$

while

$$\Delta R = \frac{\Delta R}{R} R \approx 0.05 \times 6.5 \text{ ohm} \cong 0.3 \text{ ohm}.$$

Thus, both the first and second methods give the same answer for this example.

The neglect of second and higher order terms in eq. (10) is strictly valid only when both $\Delta X/X$ and $\Delta Y/Y$ are small. In practice, you are usually only interested in a rough estimate of $\Delta F/F$, so eq. (12) will give satisfactory results as long as $\Delta X/X$ and $\Delta Y/Y$ are each a few percent or less. But if $\Delta Y/Y$ or $\Delta X/X$ (or both) become large (say 30%), you might have to use the first method to obtain satisfactory results. This is especially true in cases where a measured quantity is raised to a high power in the calculation of F.

2. <u>Examples</u>

The remainder of this section will show how eq. (12) reduces in practice to some of the more common types of functions.

Sums

$$F = B(X + Y) \qquad B = \text{constant}$$

$$\Delta F = B(\Delta X + \Delta Y)$$

$$\frac{\Delta F}{F} = \frac{\Delta X + \Delta Y}{(X + Y)}$$

<u>Example: Error in an average</u>. First, suppose that you measure the diameter D of a wire twice in a row, using the same micrometer both times. The results (D_1 and D_2) will differ

slightly with each setting, but the estimated bounds of error will be the same for each measurement (you are measuring the same thing with the same instrument each time). You might obtain

$$D_1 = 0.323 \pm 0.005 \text{ mm}$$

$$D_2 = 0.326 \pm 0.005 \text{ mm}.$$

You might decide to use the average $(\overline{D})$ of D_1 and D_2 to characterize the diameter of the wire:

$$\overline{D} \equiv \tfrac{1}{2}(D_1 + D_2)$$

$$\Delta\overline{D} = \tfrac{1}{2}(\Delta D_1 + \Delta D_2) = \tfrac{1}{2}(0.005\text{mm} + 0.005\text{mm}) = 0.005\text{mm}.$$

Note that when all the data is measured to the same accuracy, then the accuracy of the average value is the same as the accuracy of an individual measurement. (This is strictly true only for cases where we have just a few measurements: in the case where we have several hundred measurements of the same quantity with the same instrument, then the accuracy of the average may be higher than the accuracy of an individual measurement; because when so many values of the same measurement are added together, the indeterminate errors will probably cancel each other to a significant extent, leaving only the systematic errors.

Next, suppose that you again measure the diameter of a wire but now under different circumstances: this time you make the first measurement with one micrometer and the second measurement with another micrometer. Suppose that the second micrometer is much less accurate than the first. Your measurements might be

$$D_1 = 0.325 \pm 0.005\text{mm}$$

$$D_2 = 0.32 \pm 0.03 \text{ mm}.$$

If you decide to use the average of these two measurements to characterize the diameter of the wire, then you would have to quote an error of

$$\Delta\overline{D} = \tfrac{1}{2}(\Delta D_1 + \Delta D_2) = \tfrac{1}{2}(0.005\text{mm} + 0.03\text{mm}) \approx 0.02\text{mm}.$$

In this case, the error in the average is four times larger than the error in your most accurate measurement $(\Delta\overline{D} \approx 4\Delta D_1)$. Thus, in a case like this one, where the accuracies of the measurements differ greatly, nothing is to be gained by taking the average. You would be better off simply using the result of the most accurate measurement.

Differences

$$F = B(X - Y) \quad \quad B = \text{constant}$$

$$\Delta F = B(\Delta X + \Delta Y)$$

$$\frac{\Delta F}{F} = \frac{\Delta X + \Delta Y}{X - Y}.$$

(Note that the errors ΔX and ΔY are added.)

Many of the practical conclusions to be drawn from the latter expression have already been discussed above (Errors in Direct Measurements). One of the main lessons is the following: if a result is determined by taking the difference of two nearly equal quantities X and Y, both X and Y must be measured very accurately in order to obtain any reasonable relative accuracy in the result X - Y.

Products and Quotients

1. <u>Direct Proportion</u>

$$F = BX \quad \quad B = \text{constant}$$

Then $\Delta B = 0$, and

$$\Delta F = B \Delta X$$

$$\frac{\Delta F}{F} = \frac{\Delta X}{X}.$$

The fractional error in F is the same as the fractional error in X, but the <u>absolute</u> error in F is <u>B times</u> the absolute error in X. Similar statements can be made about the role of the constant B in the other equations above and below.

Example: circumference C of a circular disc of diameter D. We have

$$C = \pi D \approx 3.14 D$$

$$\Delta C = 3.14 \, \Delta D$$

$$\frac{\Delta C}{C} = \frac{\Delta D}{D}.$$

If D is measured to ±1 mm then the result for C is only good to about ±3mm.

2. General Case

The previous example considered a special case of a more general rule. First, note that any product or quotient can be written as a product of terms, each term raised to some power:

$$F = BX^a Y^b \qquad B = \text{constant}$$

Here the exponents a and b are constant real numbers. They may be either positive or negative and also need not be integers. Then the error in F is

$$\Delta F = \left| B(aX^{a-1})Y^b \cdot \Delta X \right| + \left| BX^a(bY^{b-1}) \cdot \Delta Y \right|$$

with

$$\frac{\Delta F}{F} = \left| a \frac{\Delta X}{X} \right| + \left| b \frac{\Delta Y}{Y} \right| .$$

If X and Y are positive this can be more compactly written as

$$\frac{\Delta F}{F} = |a| \frac{\Delta X}{X} + |b| \frac{\Delta Y}{Y} .$$

Example: The density (d) of a solid material can be determined by fabricating a solid cylinder out of the material, measuring the mass (M), diameter (D), and length (H) of the cylinder and then calculating d from the equation.

$$d = \frac{M}{V} = \frac{M}{\frac{\pi}{4} D^2 H} = \frac{4}{\pi} M^{+1} D^{-2} H^{-1}$$

Then from the above considerations

$$\frac{\Delta d}{d} = |+1| \frac{\Delta M}{M} + |-2| \frac{\Delta D}{D} + |-1| \frac{\Delta H}{H}$$

$$= \frac{\Delta M}{M} + 2 \frac{\Delta D}{D} + \frac{\Delta H}{H}$$

and

$$\Delta d = \left(\frac{\Delta M}{M} + 2\frac{\Delta D}{D} + \frac{\Delta H}{H}\right) \cdot d \quad .$$

Note that raising a measured value X to some power magnifies the effect that ΔX has on the accuracy of the calculated result. Thus, numbers that are to be raised to higher powers in the course of a calculation must be measured with particular care.

Example: The cross sectional area (A) of a metal wire of circular cross section can be calculated from the diameter (D) of the wire using the equation

$$A = \frac{\pi}{4} D^2$$

then

$$\frac{\Delta A}{A} = 2 \frac{\Delta D}{D} .$$

Here the relative error in the result is twice the relative error in the measured quantity; if $\frac{\Delta D}{D} = 3\%$ then $\frac{\Delta A}{A} = 2 \times 3\% = 6\%$.

Example: If a mass (M) suspended on the end of a spring is set vibrating, it will vibrate with a period (T) given by

$$T = 2\pi \sqrt{M/K} = 2\pi M^{\frac{1}{2}} K^{-\frac{1}{2}}$$

Here the spring "constant" K should be considered as a "variable" in the sense that it does contribute to ΔT: K has a definite value for a given spring, but this value of K must be measured experimentally and will thus be uncertain by some amount ΔK. In contrast, the constants "2" and "π" are "exact" numbers, having definite, exact, pre-assigned values. Thus,

$$\frac{\Delta T}{T} = \frac{1}{2} \frac{\Delta M}{M} + \frac{1}{2} \frac{\Delta K}{K} \quad .$$

Example: The wavelength (λ) of light emerging from a diffraction grating is related to the grating spacing (D) and the angular position (θ) of the n-th principle maximum by the equation

$$\lambda = \frac{D}{n}\sin\theta \qquad n = \text{positive integer}$$

$$\Delta\lambda = \left|\left(\frac{d\lambda}{dD}\right)_{\text{constant }\theta}\Delta D\right| + \left|\left(\frac{d\lambda}{d\theta}\right)_{\text{constant }D}\Delta\theta\right|$$

$$= \left[\frac{1}{n}\sin\theta\right]\Delta D + \left[\frac{D}{n}\cos\theta\right]\Delta\theta$$

$$\frac{\Delta\lambda}{\lambda} = \frac{\Delta D}{D} + \frac{\cos\theta}{\sin\theta}\Delta\theta$$

(Note that $\Delta\theta$ must be expressed in <u>radians</u> not degrees.) D and ΔD will be constant for a given grating. If $\Delta\theta$ is also constant, then $\Delta\lambda$ will be smallest for the higher-order maxima (n and θ large).

CONCLUSION

You should now have the knowledge to adequately understand the error in your experiments and be able to quantitatively determine the error and uncertainty in your measurements.

In every experiment, be aware of where the greatest source of error occurs, and try to eliminate the error as much as possible. Record the uncertainty of each measurement, and determine the uncertainty of each calculation using the principles and equations outlined in this section. A summary of the equations used for error analysis is provided for reference during your write-up of the experiments. Learn these principles well for they will be useful in all subsequent laboratory work.

GRAPHICAL PRESENTATION AND INTERPRETATION OF DATA

Graphical Presentation

When an experiment is performed to determine the dependence of one variable upon another, it is customary to vary the independent variable and measure the resulting values of the dependent variable. From this experiment, a table of data is obtained. Using this data, one hopes to determine an equation which expresses the mathematical relationship of the variables and quantitatively reproduces the table of data within the limits of the uncertainty.

The method of determining the empirical equation from the data is graphical. Therefore, it is imperative that you follow the conventional methods for creating and using a graph.

1. <u>Axes</u>: The dependent variable is plotted as the ordinate (vertical axis), and the independent variable is plotted as the abscissa (horizontal axis). The independent variable is usually regarded as the variable controlled by the experimenter. Also, it is typical to refer to a graph as, y vs. x (for example) where the independent variable is x and the dependent variable is y.

2. <u>Scales</u>: The numerical scales should be chosen so that any point on the resulting curve can be quickly and easily read. This means that the scale divisions should be assigned values such as 1, 2, 5, or 10, since other values may lead to difficulty in plotting and reading the data. If possible, the smallest scale division should approximately correspond to the experimental uncertainty in the data. (It is of no use to plot a graph to 4 significant figures if the data only has 2 significant figures. On the other hand, if the data has more significant figures than the graph, one has thrown away an important part of the data.) Finally, the resulting graph should in general occupy most of the sheet of paper. You should try to maintain a sensible balance of these three requirements.

3. <u>Data Points</u>: For easy identification each data point might be surrounded by a small circle, square, etc., with error bars indicating the size of the uncertainty.

4. <u>Labels</u>: Each scale should have a clear label. The label consists of the symbol for the quantity being plotted and the units of the quantity. For exceptionally large or small values one can use powers of ten times the conventional units. This will permit the use of one or two digits in identifying a main scale division. For example, you can plot capacitance in terms of 1, 2, ... microfarads rather than plotting it as 0.000001, 0.000002, ... Farads.

5. <u>Title</u>: Each graph should have a title located at the top of the paper which briefly describes what the graph illustrates. For example, "Variation of Displacement with Elapsed Time" might be an appropriate title for a freely falling body experiment.

Following these guidelines, your graph will represent a clear and complete illustration of your experimental results.

Curve Fitting

Curve fitting is basically the process of matching an equation to a graph. For the most part, you will only be plotting straight lines, whether on linear graph paper, log-log paper, or semi-log paper. The value of a straight line plot is that constants in the empirical equation are easily determined from the slope and the intercept of the line. The following is a concise summary of some linear forms, the plot, the slope, and the intercept.

FORM	PLOT (for a straight line)	SLOPE	INTERCEPT
$y = mx + b$	y vs. x on linear paper	m	b
$y^2 = mx + b$	y^2 vs. x on linear paper	m	b
$xy = m$	y vs. $1/x$ on linear paper	m	0
$y = bx^m$	1. log(y) vs. log(x) on linear paper 2. y vs. x on log paper	m	log(b)
$y = b(10)^{mx}$	1. log(y) vs. x on linear paper 2. y vs. x on semi-log paper	m	log(b)

It may be useful to remember that $\ln(N) = 2.303 \log(N)$ for some of your plots of logarithms. In addition, remember that for log-log plots the intercept is read at x=1. The slope for a log-log plot is

$$m = \frac{\log(y_2 / y_1)}{\log(x_2 / x_1)} . \tag{1}$$

The slope for a semi-log plot is

$$m = \frac{\log(y_2 / y_1)}{x_2 - x_1} . \tag{2}$$

Finally, as you fit straight lines on log or semi-log paper, notice that the uncertainties associated with the data points will be distorted.

It is also useful to know how to calculate the uncertainty in a slope determination, since the slope is often used to determine some quantity. For example, the slope of s(t) vs. t^2 in Experiment 1, would yield g/2, if v_0 were truly zero. Since there is always some scattering in the data, and indeed v_0 may have some small but unknown value, the "best straight line" will always have some uncertainty.

For a linear graph of y vs. x, the slope is given by

$$m = \frac{y_2 - y_1}{x_2 - x_1} \quad . \tag{3}$$

The uncertainty in the slope can be found using a least squares regression analysis but that can be tedious. So we will use a simpler method using the uncertainy relations for sums and quotients, and eq. (3). This method gives

$$\frac{\Delta m}{m} = \frac{\Delta y_2 + \Delta y_1}{y_2 - y_1} + \frac{\Delta x_2 + \Delta x_1}{x_2 - x_1} \quad .$$

The above information should provide you with the knowledge necessary to create a clear, readable graph and to properly analyze the graph for the useful information it contains.

PHYSICS 202

EXPERIMENTS

EXPERIMENT 1

MOTION OF A FREELY FALLING BODY

OBJECTIVE:

To determine the acceleration due to gravity by measuring some distances and times for a freely falling body.

EQUIPMENT:

Behr free-fall apparatus (See fig. 1); DC power supply; spark generator with repetition rate 60 times per second; SPST switch; meterstick

INTRODUCTION:

In this experiment you will make measurements of distance and time for a freely falling body in order to verify theoretical predictions and to verify the value of g, the acceleration due to gravity. You will analyze the data to determine a functional relationship of distance vs. time and of velocity vs. time, and check this with what you would expect from the theoretical equations. This method is used extensively in scientific work.

The equation of distance as a function of time for a freely falling object is described, according to theory, through the equation

$$s(t) = v_0 t + \tfrac{1}{2} g t^2 \,. \tag{1}$$

In additon, the equation for the velocity of the object is

$$v = v_0 + gt \,. \tag{2}$$

As seen from the functional dependence, if distance vs. time is plotted, the graph is a parabola. If velocity vs. time is plotted, the graph is a straight line with a slope of g and an intercept of v_0.

PROCEDURE:

1. Place the tape between the wires, and with the metal object at the top (held by the electromagnet) turn the sparking switch of the spark generator on.

************Caution**************************High Voltage*****************
The spark generator is a high-voltage device that can give unpleasant shocks. Be careful when using it. Do not touch the generator. Depress the sparking switch with the end of a wooden meter stick.
************Caution**************************High Voltage*****************

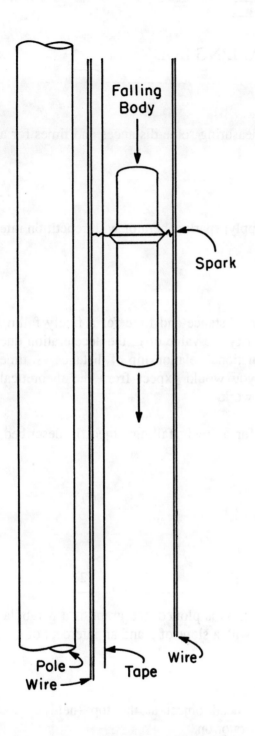

Fig. 1. Behr free fall apparatus.

2. Release the object by turning off the electromagnet.

3. Without scratching the tape, remove it for analysis. Examine the tape for any missing spots, and repeat step (2) if necessary.

4. Secure the waxed tape to the table with masking tape, with the top of the waxed tape at the left. Starting at the right where the dots are widely spaced, circle every third dot until you have 9 or 10 circled dots. Count missing dots as if they were present. Beginning with the second lowest dot, draw a square around every third dot. Thus you will have two sets of marked dots and a third set of dots which are unmarked. Examine each of the sets and choose the best set with no missing dots for further analysis. See fig. 2.

5. Prepare a data table similar to table 1.

6. Beginning at the left end of the tape where the dots are closely spaced, lay a meter stick on edge as close as possible to the line of dots. Record the position of the dots $s(t_n)$ in cm, estimating to two decimal places. Record the time t_n corresponding to each $s(t_n)$. Each (third) dot is 3/60 or 0.05 sec apart.

7. Calculate Δs_n, the distance of fall during each time interval, by subtracting successive measurements of $s(t_n)$.

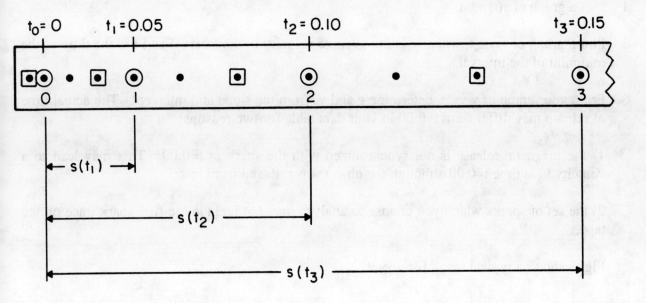

Fig. 2. Tape measurements

8. Calculate the average velocity v_n for each interval by dividing Δs_n by 0.05 sec, the time elapsed. Record the time of the midpoint of the interval (i.e., 0.025, 0.05, 0.10, etc.).

9. Check that the average velocities increase uniformly. If they do not, you should review your calculations for errors.

Dot Number n	t_n (sec)	$s(t_n)$ (cm)	Δs_n (cm)	v_n (cm/sec) midpoint	t(sec) midpoint
0	0.00	0.00	0.00		
1					
2					
3					
4					
5					
6					
7					
8					
9					

Table 1. Format for data table.

ANALYSIS:

1. Plot a graph of s(t) vs. t.

2. Plot a graph of v vs. t, where v is the average velocity in the interval and t is the time of the midpoint of the interval.

3. From your graph of v vs. t, determine g and v_0 from the slope and intercept. The actual time of release may differ from t=0.00 in your data table for two reasons:

   1) the magnetic release is not synchronized with the spark at t=0.00. This may lead to a velocity v_0 at time t=0.00 although the object was released from rest.

   2) the set of sparks which you choose to analyze may not include the first spark trace on the tape.

Fig. 3 shows a typical result for v vs. t.

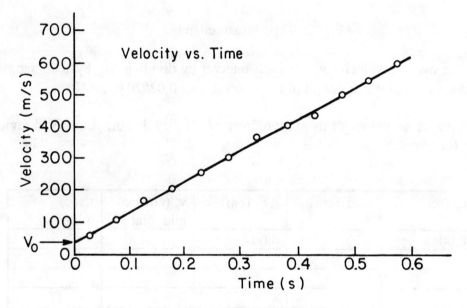

Fig. 3. Typical result for v vs. t curve.

4. Write equations similar to eq.(1) and eq.(2) using values determined from your data. Calculate $s(t_n)$ from these equations and compare these values to the measured value.

5. Suppose you hold an object motionless about 4 ft. above the ground and then let it fall to the ground without interference. About how long does it take to hit the ground? (Use your equations and compare it to the value found from your s vs. t graph.)

EXPERIMENT 2

EXPERIMENTAL ERROR - PENDULUM

OBJECTIVE:

To verify the theoretical equation for the period of a simple pendulum using direct measurement and error analysis.

EQUIPMENT:

Meterstick; sturdy (non-stretching) thread; pendulum bob; your wristwatch.

INTRODUCTION:

The equation for the period of a simple pendulum is

$$T = 2\pi\sqrt{\frac{\ell}{g}} \qquad (1)$$

where ℓ is the distance from the point of support of the pendulum to the center of mass of the pendulum bob, and g is the acceleration due to gravity. Note that the period is independent of mass.

To verify eq. (1), you can measure T directly and compare it to the value of T calculated using eq. (1). You measure T directly by setting the pendulum in motion and using a watch to time the oscillations. Let us call this measured value T_{MEAS}. You calculate T by measuring the length of the pendulum, knowing the value g = 980.120±0.005 cm/s^2 for State College and using eq. (1). Let us call this predicted value T_{CALC}.

As discussed in the section entitled "Experimental Errors", experimental measurements are never ideally accurate. Because T_{MEAS} and T_{CALC} will seldom be equal, it is necessary to see whether the random errors in the measurements are sufficient to explain the discrepancy. The random errors in T_{MEAS} and T_{CALC} can be calculated using the appropriate relations for error analysis. If T_{MEAS} and T_{CALC} are equal within the range of experimental uncertainty, you can feel reasonably certain that eq. (1) is a correct description of the system.

If your values of T do not agree after including the effects of experimental errors, you can proceed as follows:

1. Check your calculations for arithmetical mistakes.

2. Review your determination of ΔL. Were you unreasonably optimistic in your estimates of measurement accuracy? Are your instruments calibrated as well as you thought, and how precisely can you read them?

3. Repeat the measurements.

4. Reconsider the experimental situation. Usually a number of simplifying assumptions are made when deriving an equation. Does your experimental setup satisfy all these assumptions? If not, would the resulting effect be large enough to explain the discrepancy in your results?

PREPARATION:

Know how to calculate the absolute and relative errors for L, T_{MEAS} and T_{CALC}. If you do not have the equations already prepared, you may not finish the report in time. Refer to "Summary of the Propagation of Indeterminate Errors" for easy reference.

PROCEDURE:

1) Construct a pendulum about 30 cm long. The pivot point should be as frictionless as possible.

2) Make a table similar to Table 1 to record the data necessary to determine T_{CALC}.

R_s	±
$R_b =$	±
$L =$	±
$g = 980.12 \pm 0.005$ cm/sec	
T_{CALC}	±
$\Delta T_{CALC} =$	±
$\Delta T_{CALC} / T_{CALC} = \pm$	

Table 1. Determination of T_{CALC}

3) Measure the length of the pendulum. Record R_s, the scale reading at the point of support, and R_b, the scale reading at the center of mass of the bob; and determine L according to

$$L = R_b - R_s \quad . \tag{2}$$

4) Estimate the uncertainties in the scale readings. Note that the uncertainty in R_b will be different from the uncertainty in R_s, since the position of the center of mass is not as easily measured as the position of the point of support.

5) Make a table similar to table 2 to record the data necessary for determining T_{MEAS}.

X_i	X_f	n	t	Δt	$\Delta t / t$	T_{MEAS}	ΔT_{MEAS}	$\Delta T_{MEAS}/T_{MEAS}$
		5						
		100						

Table 2. Determination of T_{MEAS}

6) Displace the pendulum bob sideways about 2 cm from its equilibrium position and then release. Let it swing back and forth a couple of times. When the pendulum reaches a turning point, count "zero" and start timing. Record the intial time, X_i. When the pendulum returns to the starting point, count "one". It has completed one cycle. Continue for 5 cycles, and stop timing on the count of "five". Record the final time, X_f. Remember to record the uncertainties of the time measurements.

7) Repeat for 100 cycles, and record the initial and final times. Note that eq. (1) is valid only for small amplitudes.

8) Estimate the uncertainty in the time measurements.

ANALYSIS:

For the following steps, be sure to show all your work clearly in the "calculations" section of your lab report. For each quantity calculated, give the algebraic formula and then show all steps in the numerical calculation.

1) Calculate ΔL, assuming that the meter stick calibration is correct to 0.05% for distances comparable to 30cm. You may want to refer to the "Summary of the Propagation of Indeterminate Errors" for the appropriate equation, remembering to include the calibration error by treating it as a random error.

2) Calculate T_{CALC} from Eq.(1). For State College g = 980.120 ± 0.005 cm/sec².

3) Determine the relative error and the absolute error for T_{CALC}.

4) Calculate T_{MEAS} for each number of cycles. Then, let t = the time for n cycles

$$T_{MEAS} = t / n \qquad (3)$$

5) Determine the relative error and the absolute error for T_{MEAS}. (The calibration error of a watch is quite negligable.)

6) State clearly how T_{MEAS} and T_{CALC} overlap for each of your determinations of T_{MEAS}. That is, verify eq. (1) from your results.

7) Which value of T_{MEAS} is closer to T_{CALC}? Is this what you would expect from your knowledge of experimental errors?

EXPERIMENT 3

EXPERIMENTAL ERRORS - VTVM

OBJECTIVE:

To use a vacuum tube voltmeter (VTVM) to make voltage measurements in a simple circuit and to learn how to calculate experimental error.

EQUIPMENT:

5V power source; a 1000 Ω resistor and a 180 Ω resistor; knife switch; VTVM

INTRODUCTION:

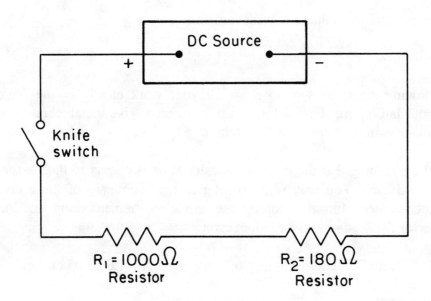

Fig. 1: Simple circuit

In this experiment you will learn how to use a VTVM and how to determine the experimental error in reading the scales. You will set up the circuit shown in fig. 1, and use the VTVM to measure the voltage across each resistor. You will analyze the data by calculating the expected voltages and comparing them with the measured values. In addition you will make measurements on different voltage scales to show that they have different accuracies. You will need to determine the experimental error in all of your measurements.

The following is a guideline to help you determine the experimental error for a particular scale.

> Voltage per division =
> Uncertainty in zero-position, ΔV_0 =
> Uncertainty in scale-reading, ΔV_s =
> Total Uncertainty, $\Delta V = \Delta V_0 + \Delta V_s + \Delta C(V_s/V_f)$, =
> where ΔC is the calibration error of full scale and V_f is the full-scale reading

Table 1: Experimental error calculation.

PREPARATION:

From Ohm's Law, V=IR, calculate the voltages that you would expect across each resistor. The power supply is 5V ± 1%.

PROCEDURE:

1. Make a simple freehand sketch of the VTVM. After you have made your sketch, label the following parts. 1) All positions of FUNCTION switch, 2) all positions of RANGE switch, 3) each of the three probes, 4) all scales on the meter face (for each scale, label the zero, mid-scale value, full-scale value, name of scale), 5) zero-adjust knob, 6) ohms-adjust knob, 7) pilot light, 8) AC supply line. Making and labeling the sketch should take you less than 10 minutes.

   Keep in mind that this experience is intended to help you become familiar with the VTVM so that you can use it with greater understanding, speed, and self-assurance. The VTVM has many component pieces, knobs, switches, scales, markings. Making a drawing helps you to become familiar with the instrument in that the action of drawing the item helps fix it permanently in your memory. You will find that this technique of sketching new instruments proves helpful throughout this course and in laboratory work in general. Hand in your drawing <u>before</u> setting up the apparatus for this experiment. Your instructor will check it for you right away.

2. Prepare a data table similar to table 2 to record the voltage range, the voltage across each resistor, the absolute error, and the relative error.

Range (volts)	V_1 (volts)	ΔV_1 (volts)	$\Delta V_1/V_2$	V_2 (volts)	ΔV_2 (volts)	$\Delta V_2/V_2$	V_2/V_1	$\Delta(V_2/V_1)/V_2/V_1$
15V								
5V								
1.5V								

Table 2: Voltage measurements with VTVM

3. Turn on the VTVM to allow it to warm up.

4. Set up the circuit of fig. 1 with the power supply on and the knife switch closed.

5. Set the "FUNCTION" switch to "DC+."

6. Set the "RANGE" to 15 V.

7. Adjust the zero reading if necessary.

8. Measure the voltage across R_1 and the voltage across R_2 (Remember to connect the VTVM in parallel with the resistor).

9. Determine the scale-reading uncertainty and the zero-reading uncertainty for your measurements.

10. Repeat steps (6) - (8) for the 5 V range and the 1.5 V range.

ANALYSIS:

1. Calculate ΔV and $\Delta V/V$ for each measurement. The manufacturer quotes a calibration error of ±3% of full scale for this VTVM.

2. Compare the expected voltages (from the PREPARATION) to the values obtained on the most sensitive scale. In each case the relative uncertainty in the power supply voltage is 1%.

3. Compare V_2/V_1 to the ratio of resistance R_2/R_1 for the 5V range. Note, in determining V_2/V_1, that the 3% calibration error and the 1% source voltage uncertainty play no role. Hence, the relative error in V_2/V_1 should be quite small.

EXPERIMENT 4
ELECTRIC FIELD PLOTTING

OBJECTIVE:

To plot the equipotential lines in two dimensions due to two equal and opposite point charges.

EQUIPMENT:

25 V transformer; DMM; pan filled with water; plastic graph sheet; electrodes.

INTRODUCTION:

The electric field due to a source charge (or charges) can be understood from Coulomb's Law; it is the force on a test particle divided by the charge of the test particle. Since this implies both magnitude and direction, electric field is a vector quantity. It can be represented by lines of force constructed such that the electric field at a point is in a direction tangent to the line of force through that point. The electric field is **always** in the direction of the force on a positively charged test particle.

Since there is a force acting on a test charge, there is a potential energy associated with the test charge at different points displaced from the source. The potential energy of the test particle divided by its charge is known as the electric potential.

In an electric field, there are many points with the same potential. The locus of such points in two dimensions is called an equipotential line. It takes no work to move a test particle along an equipotential line because its potential energy stays constant. Thus, there can be no component of electrical force along an equipotential, and the lines of force (and electric field) will be perpendicular to the equipotential lines, as in fig. 1.

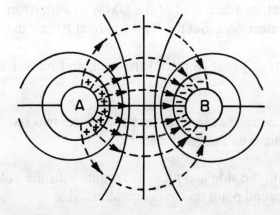

Fig.1. The equipotential lines (solid) and the electric field lines (dashed) of two equal and opposite charges.

In this experiment you will use the apparatus shown in fig. 2 to find the equipotential lines of two equal but oppositely charged bodies. Using the equipotential lines, you can draw the electric field lines.

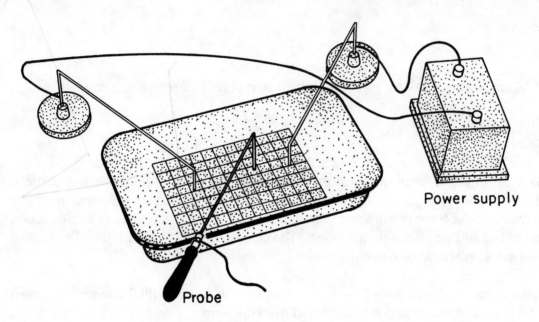

Fig. 2: Apparatus used to find the equipotential lines.

PROCEDURE:

1. Set up the apparatus of fig. 2. Position the fixed electrodes at main intersections on the x-axis of the graph sheet, and connect each electrode to one terminal of the transformer. Connect the black lead of the DMM to a transformer terminal and the red lead to the movable electrode. Set the DMM to read AC on the 200 V scale.

2. Obtain a piece of graph paper to record the data, and mark the positions of the fixed electrodes.

3. Place the movable electrode at some point between the two fixed electrodes, and record its position and potential on the graph paper.

4. Place it about 1 cm to the side and move it around until the potential is the same as in step (3). This is a second point on the equipotential line.

5. Continue with step (4) until enough points are found and an equipotential line can be drawn on the graph paper.

6. Place the electrode between the two fixed electrodes, but several centimeters from the first position. Find another equipotential line.

7. Continue with this process until the entire region has been "mapped out."

8. Plot the electric field such that it is everywhere perpendicular to the equipotential lines. Your results should resemble fig. 1.

ANALYSIS:

1. In your own words, explain why the equipotential lines are perpendicular to the electric field. You may want to mention something about conservative fields, or use the relations $E_x = -\frac{dV}{dx}$, etc., for y, z components, or $\Delta V = -\int \mathbf{E} \cdot d\mathbf{s}$.

2. What can you say about the electric field at the surface of a perfect conductor? What about the electric field inside a perfect conductor.

3. Given conductors as shown on the following page. Draw schematic electric field lines and equipotentials, and include this with your lab report.

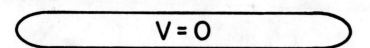

Fig. 3. Sample electrodes.

EXPERIMENT 5

D. C. CIRCUITS, PART I: COMPONENTS

OBJECTIVE:

To study the relationship between current and voltage for a carbon resistor and for a silicon diode.

EQUIPMENT:

5V power supply; switch and fuse; rheostat; ammeter; voltmeter; carbon resistor; silicon diode.

INTRODUCTION:

In this experiment you will be studying the relationship between current and voltage for a sample by determining its current-voltage characteristic curve, otherwise called an I-V characteristic. The current that you will be working with is direct current, current of constant magnitude which always flows in the same direction.

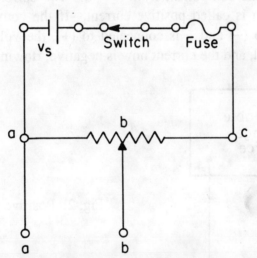

Fig. 1. Voltage-divider circuit

To measure current as a function of voltage, you will need to construct the adjustable voltage source shown in fig.1. The components of the voltage source will consist of a 5V power supply, a switch with a fuse, and a rheostat. The main element of a rheostat is a long piece of resistance wire wrapped in a spiral onto an insulating tube. By coiling the wire in this way, the long wire can be contained in a small space. Another part of the rheostat is a sliding contact that can move along the resistance wire, allowing one to make electrical contact with various points on the wire. In fig. 1 the arrowhead represents this sliding contact. By choosing the contact point, the voltage V_{ab} can be adjusted to any value between 0 and the source voltage V_s. As you can see, when b coincides with a, $V_{ab} = 0$. As b moves toward c, V_{ab} increases; and when b coincides with c, $V_{ab} = V_{ac} = V_s$. The circuit of fig. 1 is called a voltage-divider circuit. It divides the source voltage V_s into two parts, V_{ab} and V_{bc}, and makes one of the parts, V_{ab}, available at the output terminals a,b. The fuse protects the battery against the excessive current drain that would result from an accidental short circuit: if too much current flows through the fuse, the fuse wire melts; this breaks the conducting path and prevents futher current flow from the battery.

To determine the I-V characteristic, an adjustable voltage is applied across the terminals of a component. The voltage across the component and the current through it are then measured. Next, the voltage is adjusted, and the new voltage and resulting current are measured. The effects of both positive and negative voltages are considered.

Negative voltages and negative currents can be understood by knowing how they are defined. Consider a sample, S, in a circuit such that terminal a is connected to the low potential (−) terminal of the voltage source, and terminal b is connected to the high potential (+) terminal of the voltage source, as in fig. 2. This can be arbitrarily defined as positive voltage across the component, which is said to be forward biased. The current always flows from the (+) terminal of the source to the (−) terminal; therefore, the current flows through the component from b to a. The current flowing across the component in this direction is called positive current. If the component is "reversed" such that b is connected to (−) and a is connected to (+), the voltage is negative, the component is reverse biased, and the current now is negative, flowing from a to b.

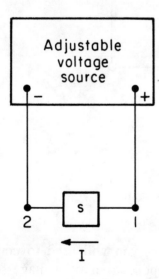

Fig. 2. Positive voltage

The I-V characteristic for a short tungsten wire (the filament of a light bulb) is shown in fig. 3. Note that it is symmetric about the origin since $I(-V) = -I(V)$. A symmetric I-V characteristic indicates that +V and −V produce the same magnitude of current.

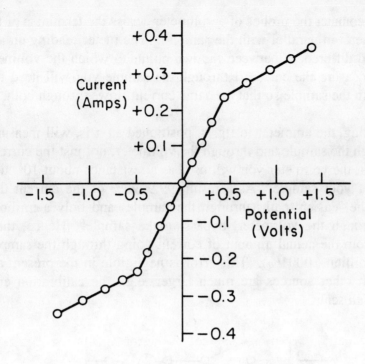

Fig. 3. Symmetric I-V characteristic.

The resistance of a substance having a symmetric I-V curve is defined as

$$R = \frac{V}{I} \quad (1)$$

where R is measured in Ohms, V is measured in Volts, and I is measured in Amps. For an ideal resistor, R does not depend on the current and, therefore, obeys Ohm's Law. Materials with this property are said to be ohmic. For other resistive devices, R is still defined by eq.(1); however, it varies as the current (and voltage) change. In practical terms, the resistance of ohmic substances may be thought of as the potential difference required to produce a current of one Ampere.

Consider eq.(1) for an ohmic material. In this case R is constant and

$$I = \frac{V}{R}.$$

If current I is plotted as a function of voltage V, the curve will be linear with slope 1/R.

In this experiment you will set up the circuit in fig. 4 in order to apply an adjustable voltage across each sample, S, and measure the voltage across it and the current through it. From your data, you can construct an I-V characteristic for the device and (if the I-V curve is symmetric) determine its resistance R. To measure the voltage

V_{ab} you will need to connect the probes of a voltmeter across the terminals of the sample (i.e., connect a voltmeter in parallel with the sample). The meter reading on a voltmeter indicates the potential differences between the two points to which the voltmeter probes are connected. To measure the current I through the sample you will need to place an ammeter in series with the sample so that the same current flows through both.

Strictly speaking, the ammeter in fig. 4, positioned as it is, will measure the sum of the currents through the sample and through the voltmeter, not just the current through the sample alone. But the voltmeter you will use has a resistance about 10^6 times larger than the resistance of your sample. As a result, 99.9999% of the current through the ammeter has taken the "easier" path (through the sample) and only a millionth of the current has gone through the voltmeter, bypassing the sample. Hence, the ammeter reading will differ from the actual amount of current going through the sample by only about one part in a million (.0001%). This error is negligible in the present application since the errors from other sources are much larger, e.g., the calibration error of the ammeter is 0.5% of full scale.

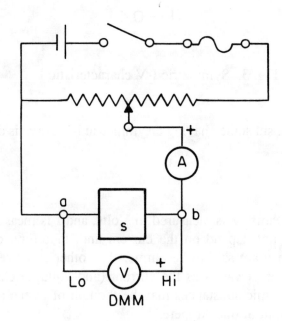

Fig. 4: Circuit for measuring I and V through the sample S.

PROCEDURE:

1. Set up the voltage-divider circuit of fig.1, but **do not connect the power supply**. Once your instructor has checked your circuit, you may connect the power supply. Measure V_{ab} with the DMM, and note how the voltage changes as you change the setting of the rheostat. **At this point, neither sample is in the circuit.** Which position of the rheostat gives you a high voltage? What about a low voltage?

2. Set up the circuit in fig. 4 using a carbon resistor as the sample, S. A is an ammeter connected in series, and V is a DMM connected in parallel, with the high (+) connected to b (+) and the low (−) connected to a (−). (It is important to remember that the voltmeter is always connected in parallel, and an ammeter is always connected in series with the sample. Therefore, a voltmeter has an extremely high resistance, thus allowing almost no current to flow through it, and an ammeter has an extremely low resistance, thus giving a very low voltage drop).

3. Set the voltage and record the values of V and I; be sure to record the uncertainty with each measurement. **Keep the values for I below 0.4 amps for the carbon resistor and 0.3 amps for the diode, to avoid permanently damaging your samples.** Take 3 data points for +V. Disconnect the sample and reconnect it in the opposite direction so that the current flows from b to a. Record 3 data points for −V. Remember that the current is positive for +V, and the current is negative for −V. You should have 6 data points, 3 for −V and 3 for +V.

4. Replace the resistor with the diode, and set up the diode such that it is forward biased. Make measurements of V and I for +V and −V. The diode does not allow current to flow in reverse bias; therefore, you should verify that for −V, no current flows through the diode.

ANALYSIS:

1. Plot I vs. V for the carbon resistor. Include error bars on the data points.

2. Determine R from the slope of the graph, and calculate ΔR, the uncertainty in R. Compare your determined value to the nominal value indicated on the resistor. (The manufacturer quotes a tolerance of ΔR/R = 10%.) You may need a chart to interpret the color code on the resistor.

3. Write the equation for the I-V characteristic of this carbon resistor.

4. Plot I vs. V for the silicon diode.

5. In the discussion section of your lab report, analyze the I-V characteristic for the resistor and the diode. Discuss the linearity, symmetry, and whether the device is ohmic or not.

6. For an ohmic device, R does not depend on voltage or current, but may not be entirely constant over long time periods. What might affect R, assuming that I remains constant?

EXPERIMENT 6

D. C. CIRCUITS, PART II: EMF, INTERNAL RESISTANCE, AND D. C. POWER

OBJECTIVE:

To measure the emf and internal resistance of a DC power source and to investigate the power output of a DC power source with various loads.

EQUIPMENT:

5V power supply; ammeter; decade resistance box; DMM.

INTRODUCTION:

A device which can maintain different levels of potential across its terminals is said to be a source of emf. Consider fig. 1. If the terminals a and b are connected to an external circuit, then the emf source ($\mathcal{E}$) as shown, drives positive charge around the circuit from point a to point b through the resistor, R. All real power sources have a finite resistance which causes the terminal potential difference V_{ab} to be different from the actual emf $\mathcal{E}$.

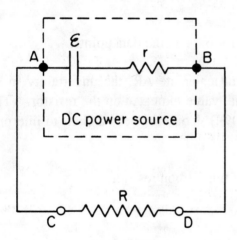

Fig. 1. Power source with load R.

For a current flowing in the circuit in fig. 1, Ohm's Law and Kirchoff's Rules give the relation between the emf $\mathcal{E}$ and current I as

$$\mathcal{E} = IR + Ir \ . \tag{1}$$

Since $V_{ab}=IR$, the terminal potential difference is

$$V_{ab} = \mathcal{E} - Ir \ . \tag{2}$$

This is the equation of a straight line. If V_{ab} vs. I is plotted as in fig. 2, the slope is $-r$ and the intercept is $\mathcal{E}$.

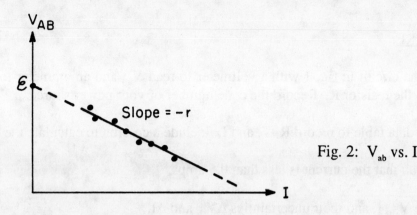

Fig. 2: V_{ab} vs. I

The current in the circuit as determined from Eq. (1) is

$$I = \frac{\mathcal{E}}{(R+r)} \tag{3}$$

The power delivered by the emf ($P=\mathcal{E}I$) must equal the power dissipated by the rest of the circuit. The power dissipated by a resistor is given by I^2R, otherwise known as the Joule heating, or I squared R loss. Therefore, the total power delivered by the emf is

$$\mathcal{E}I = I^2R + I^2r = \frac{\mathcal{E}^2}{R+r}. \tag{4}$$

The output power (the power dissipated in the load) is I^2R. If I is substituted from eq. (3), then we find

$$P_{out} = \frac{\mathcal{E}^2 R}{(R+r)^2}. \tag{5}$$

This has two interesting results. First, P_{out} varies as $\mathcal{E}^2$. Second, the maximum power delivered to the load would occur for r=0, if r could be varied. In reality, however, r does not equal 0 and is beyond our control. Assuming that $\mathcal{E}$ and r are constant, P_{out} approaches 0 as R→0, or as R →∞. The value of R giving the maximum value of P_{out} can be found by differentiating eq. (5) with respect to R and setting the result equal to 0.

$$\frac{P_{out}}{dR} = \frac{\mathcal{E}^2(r-R)}{(r+R)^3} = 0. \tag{6}$$

Clearly, the solution is R=r. Therefore, the maximum power delivered to a load occurs when the load resistance R is equal to the internal resistance r.

In this experiment you will plot the terminal voltage V_{ab} vs. the current I for the circuit in fig.1. From this plot you can determine the emf $\mathcal{E}$ and the internal resistance r for your power source. You will then determine the power output for various values of R, and make a plot of P_{out} vs. R, comparing with the predictions of eq. (5) with $\mathcal{E}$ and r as previously determined. You will locate the peak position of your experimental P_{out} vs. R curve, which will serve as an alternate means of determining r.

PROCEDURE:

1. Set up the circuit in fig. 1 with a voltmeter to read V_{ab} and an ammeter to read the current I through the resistor R. Record the code number of your power source.

2. Set up a data table to record R, V, and I. Include a column to calculate the power P.

3. Set R such that the current is less than 0.5 amps.

4. Measure V_{ab}, I, and their uncertainties ΔV_{ab} and ΔI.

5. Vary R and repeat step (4).

6. Repeat step (5) until you have about 10 data points.

ANALYSIS:

1. Plot V_{ab} vs. I as in fig.1. Draw error bars on your graph to indicate the uncertainty in the data.

2. Calculate r and $\mathcal{E}$ from the graph, and calculate their uncertainties.

3. Calculate P_{out} for each data point. $P_{out} = IV_{ab}$.

4. Plot P_{out} vs. R, and compare with that calculated from eq. (5), with $\mathcal{E}$ and r as previously obtained.

DISCUSSION:

1. In Step (4) of the Analysis, how do the positions of the peaks of the measured and theoretical curves compare? Does the peak position, as determined from experiment, give a good alternate means for measuring r?

2. The efficiency is defined as

$$\text{Eff} = \frac{P_{out}}{P_{delivered\ by\ emf}} = \frac{I^2 R}{\mathcal{E} I} = \frac{I^2 R}{I^2(R+r)} = \frac{R}{R+r}.$$

   i) Where is the efficiency greatest? Is it practical to attempt to achieve Eff$\equiv$1?
   ii) What is the efficiency at P_{max}?

EXPERIMENT 7

RESISTANCE MEASUREMENTS - SLIDE WIRE WHEATSTONE BRIDGE

OBJECTIVE:

To measure the resistance of a length of nichrome wire using the slide wire form of the Wheatstone bridge.

EQUIPMENT:

Slide wire bridge; DMM (or galvanometer G); power supply; switch; nichrome wire; micrometer caliper.

INTRODUCTION:

The Wheatstone bridge circuit is shown in fig. 1, where R_x is the unknown resistance and R_a is a variable resistor.

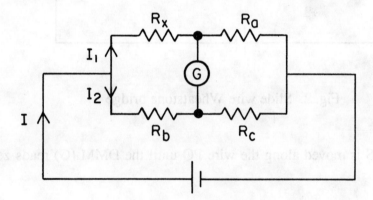

Fig.1: Wheatstone bridge

R_a is varied until no measurable current passes through the DMM, or galvanometer (G). Thus, the potential difference across R_x and R_b must be the same, so

$$I_1 R_x = I_2 R_b \qquad (1)$$

Since the current through R_a is the same as for R_x, and the current through R_c is the same as for R_b, then the potential difference across R_a and R_c must be equal, and

$$I_1 R_a = I_2 R_c \qquad (2)$$

Solving eqs. (1) and (2) for R_x, we have

$$R_x = R_a (R_b/R_c). \tag{3}$$

The apparatus shown in fig. 2 is a slide wire Wheatstone bridge. The resistance R_x is a nichrome wire of unknown resistance, and R is a decade resistance box. A long wire of uniform resistance (PQ) is stretched over a meter stick so that the length ℓ_1 has a resistance R_1 and the length ℓ_2 has a resistance R_2. Note that R_1 and R_2 formerly were called R_b and R_c.)

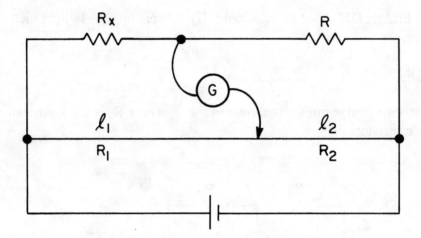

Fig. 2. Slide wire Wheatstone bridge

The DMM (G) probe S is moved along the wire PQ until the DMM (G) reads zero current. Then from eq. (3)

$$R_x = R \frac{R_1}{R_2} \tag{4}$$

Since $R_1 = \lambda \ell_1$ and $R_2 = \lambda \ell_2$, where λ is the resistance per unit length of the wire,

$$R_x = R \frac{\ell_1}{\ell_2}. \tag{5}$$

In this experiment, you will measure the resistance of the nichrome wire R_x using the bridge and calculate the resistivity of nichrome. The resistivity of nichrome is given by

$$\rho = \frac{\pi D^2 R_x}{4L} \qquad (6)$$

where L is the length and D is the diameter of the nichrome wire.

PROCEDURE:

1. Set up the circuit of fig. 2.

2. Place the sliding probe S at the 50 cm mark, and adjust R until the DMM reads very low current. This insures that the value of R is nearly R_x so that the final balance point is close to the 50 cm mark. This will decrease the uncertainty in determining R_x.

3. Slide the probe S about one mm at a time until the DMM reads zero current. The bridge is now balanced. Record R, ℓ_1, and ℓ_2.

4. Determine the distance from the balance point that the slider must be moved before the DMM shows a current. Record this distance as $\Delta\ell$, the uncertainty in the determination of the probe position. From this, estimate the relative uncertainty in ℓ_1/ℓ_2 for $\ell_1 \cong \ell_2$.

5. Measure the length L and diameter D of the nichrome wire. Also record the uncertainties in the measurements of L and D.

ANALYSIS:

1. Determine the resistance of the nichrome wire R_x from eq. (5).

2. Calculate the relative error of the resistance, $\Delta R_x/R_x$. Use $\Delta R = 10^{-3}R + 0.01\Omega$ for the uncertainty in R.

3. Determine the resistivity ρ of nichrome using eq. (6).

4. Calculate the relative error of the resistivity, $\Delta\rho/\rho$.

5. Compare your value of the resistivity to the expected value of 1.50×10^{-6} Ωm.[1]

6. Use eq. (5) to show that the relative error $\Delta R_x/R_x$ is least when the bridge is balanced at $\ell_1 = \ell_2 = 50$cm.

---

[1] R. A. Serway, Physics (Saunders College Publishing, Philadelphia, 1990) 3rd Edition, p. 747.

EXPERIMENT 8

TEMPERATURE COEFFICIENT OF RESISTANCE

OBJECTIVE:

To measure resistances, using a Wheatstone Bridge, and their corresponding temperatures to determine the temperature coefficient of resistance of copper and a thermistor.

EQUIPMENT:

Multimeter DMM (or galvanometer G with STSP switch and a 10,000 Ω resistor); box form of the Wheatstone Bridge; power supply; specially wound copper coil; thermistor; beaker; ice; bunsen burner; wire gauze; tripod; thermometer.

INTRODUCTION:

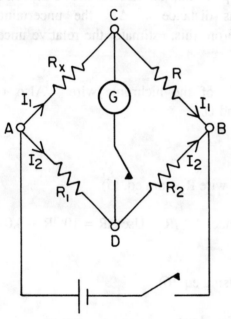

Fig. 1. Wheatstone bridge

A network of resistors, a battery, and a galvanometer arranged as in fig. 1 is known as a Wheatstone Bridge. It is used to measure resistances. In a typical experiment, one of the resistance values is the unknown to be determined. The three known resistances are varied systematically until the galvanometer G indicates no current. If, under these circumstances we apply Kirchhoff's loop rule to the closed loops ACDA and CBDC, respectively, we get

$$R_x I_1 - R_1 I_2 = 0 \tag{1}$$

and

$$R I_1 - R_2 I_2 = 0 . \tag{2}$$

By division, these two equations yield

$$\frac{R_x}{R} = \frac{R_1}{R_2} .\qquad(3)$$

The unknown resistance can be obtained using eq. (3).

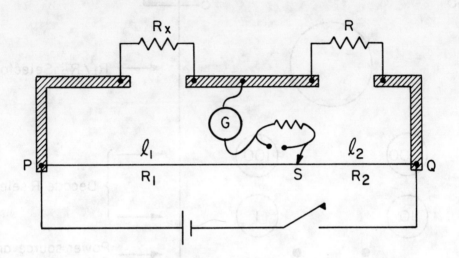

Figure 2: Slide-wire form of the Wheatstone bridge

For the slide-wire form of the Wheatstone Bridge shown in fig. 2, PQ is a wire of uniform resistance, R is a decade resistance box, and X is the unknown resistance. The galvanometer probe S is moved along the wire PQ until there is no current in the galvanometer. Then from eq. (3)

$$\frac{R_x}{\lambda(PS)} = \frac{R}{\lambda(SQ)} .\qquad(4)$$

where λ is the resistance per unit length of the wire. If PS = ℓ_1 and SQ = ℓ_2, then

$$R_x = \left(\frac{\ell_1}{\ell_2}\right) R .\qquad(5)$$

In the decade box form of the Wheatstone Bridge (fig. 3), the wire PQ of the slide-wire form is replaced by several resistors arranged circularly on a spindle. They are manipulated by a single selector switch. For each position of the switch, two of the resistors will be in adjacent-bridge arms corresponding to PS and SQ of fig. 2. The selector switch is called the multiplier or ratio arm and can give any one of the following ratios: 1:1, 1:10, 1:100, 1:1000, and their reciprocals.

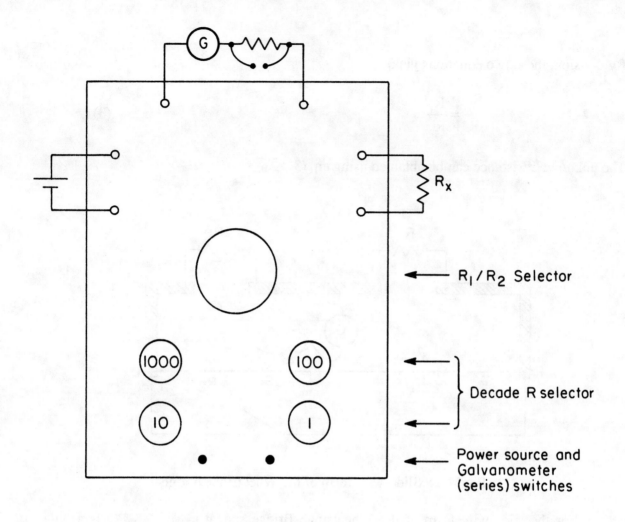

Figure 3: Decade box form of the Wheatstone bridge

The known resistance R consists of 4 decade resistors of numerical value 1, 10, 100, and 1000 ohms. Each decade has nine equal resistors. When the bridge is balanced the unknown resistance R_x is equal to the product of the ratio multiplier and the value of R. The overall limit of error of this bridge is about 0.02%R + 0.02 ohms.

Temperature Coefficient of Resistance of a Metal

The resistance of a metal varies with the temperature according to the relation

$$R_T = R_0(1+\alpha T+\beta T^2) \qquad (6)$$

where R_0 and R_T are the resistances at 0°C and T°C, respectively. α and β are constants with $\alpha \gg \beta$; therefore, over small ranges of temperature we may write

$$R_T = R_0(1+\alpha T) \qquad (7)$$

where the constant α is the temperature coefficient of resistance.

It is becoming increasingly standard to use R_{20}, the resistance at 20°C (room temperature), as a reference value. The temperature coefficient of resistance at 20°C is written α_{20}. Then Eq. (7) becomes

$$R_T = R_{20}[1+\alpha_{20}(T-20)] \quad . \tag{8}$$

One can easily determine α_{20} by using the box Wheatstone Bridge to measure the resistance of a metal at several temperatures between 0°C and 100°C, and analyzing the data graphically.

Temperature Coefficient of Resistance of a Thermistor

A thermistor is a semi-conductor of ceramic material with a high negative temperature coefficient of resistance. It is usually made of sintered metallic oxides of nickel, cobalt, iron, manganese, and a few others. The variation of resistance with temperature for a thermistor is given by the expression

$$R = R_0 e^{[(\beta/T)-(\beta/T_0)]} = R_0 e^{-\left[\frac{\beta}{T_0 T}(T-T_0)\right]} \tag{9}$$

where R is the resistance at absolute temperature T, R_0 is the resistance at reference temperature T_0 and β depends on the material used in making the thermistor. While β increases slightly with temperature, β/T may be regarded as being reasonably constant over a fairly large range of temperatures around room temperature. Thus, eq. (9) may be expressed as

$$R = R_0 e^{-\left[\frac{\beta}{T_0^2}(T-T_0)\right]} \tag{10}$$

From eq. (8), we find the temperature coefficient of resistance, expressed in derivative form, to be

$$\alpha_0 = \left(\frac{1}{R}\frac{dR}{dT}\right)_{T=T_0} \quad . \tag{11}$$

Substituting for R from eq. (10), we find

$$\alpha_0 = -\beta_0/T_0^2 \quad , \tag{12}$$

and hence

$$R = R_0 e^{\alpha_0(T-T_0)} \quad . \tag{13}$$

Let us now choose our reference temperature, again, as 20°C (293K). To find α_{20}, we express eq. (13) in logaithmic (base 10) form,

$$\log(R/R_{20}) = (T-20°C)\alpha_{20} \log(e) \quad . \tag{14}$$

In a semilog plot of (R/R_{20}) vs. T, at T=20°C the quantity (R/R_{20}) will be unity, while α_{20} can be obtained from the slope of the graph. Remember that it should have a negative value.

PROCEDURE:

1. Set up the Wheatstone Bridge, a power supply, a multimeter (or galvanometer), and the copper coil as shown in fig. 3.

2. Create a data table similar to table 1 to record the data necessary to determine the coefficient of resistance of a metal.

$R_T(\Omega)$	T(°C)

Table 1. Data for copper

3. Set the ratio multiplier of the bridge to 1. Vary R until the best balance is obtained. Set the ratio multiplier to 0.1, and vary R until the best balance is obtained. Continue to reduce the ratio multiplier, and adjust R until the value of R_T with the greatest number of significant digits is obtained.

4. Record the temperature of the room.

5. Place the copper coil into a container of melting ice until it has reached the temperature of the ice. Using the same ratio multiplier setting, balance the bridge. Record the temperature T of the copper coil and the resistance R_T.

6. Heat the ice bath and repeat step (5) at approximately 20° intervals until the boiling point of water is reached.

7. Measure the resistance of the copper coil at the temperature of liquid nitrogen. Since quick transition from the boiling water to the liquid nitrogen will damage the silicone insulation covering the coil, immerse it in ice before placing it into liquid nitrogen. The nitrogen will begin to boil as soon as the coil touches it, so wait until the boiling ceases before lowering it any further.

8. Replace the copper coil with the thermistor, and prepare a data table similar to Table 2.

R_T	R_T/R_0	T

Table 2. Data for a thermistor

9. Balance the bridge as in step (3). Record R_T and room temperature.

10. Place the thermistor in melting ice, balance the bridge, and record the resistance R_0 and the temperature T. Be as accurate as possible with this measurement since later you will be plotting R_T/R_0 vs. T.

11. Heat the ice bath and make measurements of R_T and T. Try to have approximately 4 data points below 70° and 3 data points above 70°.

ANALYSIS:

1. For the copper, plot a graph of R_T vs. T, only using the data between 0° and 100°. From the graph determine R_{20} and α_{20} and the absolute error. Remember to put Eq. (9) into the form of $y = mx + b$. (See the section on graphical analysis in the Introduction for help.)

2. Compare your result to the accepted value of α_{20} for copper, 0.0039 °C^{-1}.

3. Use your experimental values of R_{20} and α_{20} to calculate the resistance of the copper coil at −195°C, the temperature of liquid nitrogen. Compare this value with the value you obtained by direct measurement.

4. The value of α_{20} for carbon and manganin (Cu 84, Mn 12, Ni 4) are −0.0005°C^{-1} and 0.000000°C^{-1}, respectively. Of what practical significance are these values?

5. For the thermistor, plot a graph of R/R_{20} vs. T on semi-log paper (For help refer to the section on graphical analysis in the Introduction). Using the data between 0° and 70°, determine α_{20}.

EXPERIMENT 9

THE CATHODE RAY OSCILLOSCOPE

OBJECTIVE:

To become familiar with the operation and use of the oscilloscope.

EQUIPMENT:

Tektronix 2213 oscilloscope, sine/square wave signal generator, 1000 Hz oscillator, 5V power source.

> AVOID KNOB TWIDDLING OF THE OSCILLOSCOPE. FOLLOW THE INSTRUCTIONS BELOW AND MAKE AN ADJUSTMENT ON THE INSTRUMENT ONLY WHEN (a) THE INSTRUCTIONS SUGGEST IT, (b) THE INSTRUCTOR DIRECTS YOU TO, (c) THE RESULTS OF THE ADJUSTMENT ARE KNOWN BEFOREHAND, OR, (d) THE SITUATION DEMANDS IT.

INTRODUCTION

The oscilloscope is an important and versatile instrument that may be used for visual presentation of almost any kind of electrical phenomena. The indicating unit of the oscilloscope is the cathode ray tube. An electron gun at the back of the tube accelerates electrons toward the phosphor coated tube face. Focusing electrodes produce a small spot on the tube face. Since the electrons travel at high speeds, the beam adjusts very quickly to deflecting forces; thus the oscilloscope is ideally suited for displaying time-varying electrical phenomena.

In one type of oscilloscope, the deflecting forces are provided by two pairs of mutually perpendicular electrodes (called plates) across which electric fields are applied; a pair of x-plates controls the horizontal motion of the electron beam, while a pair of y-plates controls the vertical motion, as shown in fig. 1.

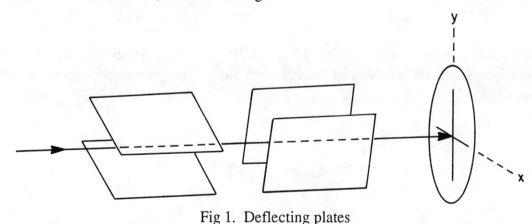

Fig 1. Deflecting plates

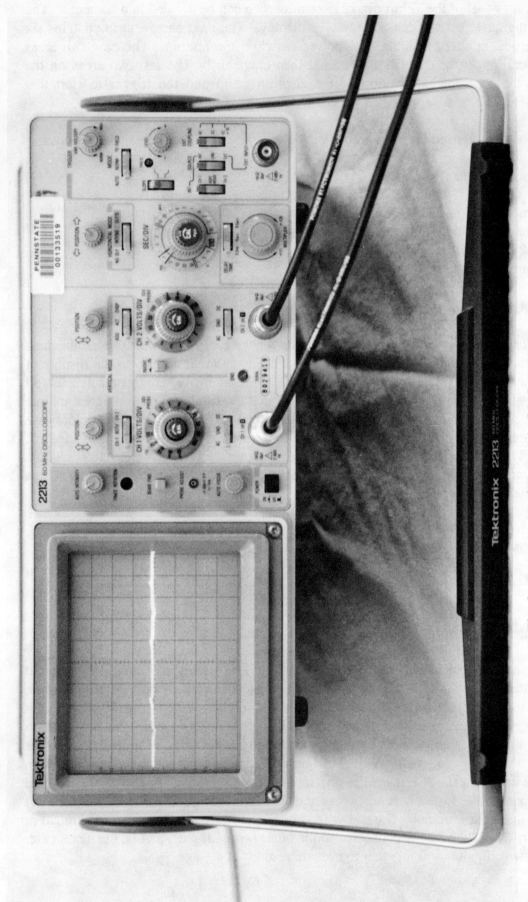

Fig. 2. Front panel of Tektronix 2213 oscilloscope.

The front panel of the Tektronix 2213 oscilloscope can be divided into six areas. The first area, to the far left is the cathode ray tube display. The next narrow strip contains six items in a vertical column, including the power switch at the bottom. The next two areas concern vertical inputs for channel 1 first, and then channel 2. The last two areas on the right control the horizontal deflection of the electron beam and the trigger to start that deflection. See fig. 2.

The controls referred to below initially should be placed in the indicated positions

1. Auto intensity: Counterclockwise.
2. Position - Channel 1, 2 and horizontal: pointing up.
3. Vertical mode switches: Ch. 1, and Alt.
4. Horizontal mode: No Dly (no delay).
5. Sec/Div: 2 ms (2 millisecond per centimeter).
6. Trigger mode: Auto (Automatic)

PROCEDURE:

Push on the power switch. Wait 20 seconds for the electron beam to be established. <u>Slowly</u> turn the "Auto Intensity" control clockwise. A line should appear on the display of the cathode ray tube. Keep the intensity low. If the "Auto Intensity" control reaches its full clockwise position and no line is found, check to see that the oscilloscope is plugged into an electrical outlet. If not plugged in, return the "Auto Intensity" control to counterclockwise position before plugging in. If it is plugged in, the electron beam is being deflected off screen. Briefly push the "Beam Find" switch (located above the power switch), to see which directions from center screen the beam is being deflected. This "Beam Find" switch compresses the display for this purpose.

> CAUTION: **Keep the intensity low to avoid damage to the phosphor coating of the tube face.**

1. <u>DC Voltage Measurements</u>

Examine the controls for vertical channel 1 (CH 1). This unit amplifies external signals admitted to the "CH 1" input, then sends them to the vertical deflection (y) plates. <u>Gently</u> turn the red variable control knob back and forth, to observe how it feels. Note that it clicks into an extreme clockwise position. Put it in this extreme clockwise "clicked" position gently. The amplifier is now on "Calibrated" and should always be used in this position for voltage measurements. Note that "Volts/Div" on the amplifier means volts per major scale division (one centimeter). A test lead is provided with a connector for the oscilloscope, a length of coaxial cable, two connecting wires and two connecting (alligator) clips. Connect this cable to the input terminal. The red wire is connected to the center (input) terminal at the connector. The black lead is connected to the outer (ground) part, that is, the metal case. Connect the two alligator clips together to provide zero input voltage.

Adjust channel 1 controls as follow:

1. Position: line in center of display
2. Vertical mode: remains at Ch 1
3. Volts/Div: 5 at the 1X side (our "probe" does not attenuate).
4. AC - GND - DC: at DC

Connect the black alligator clip to the negative (ground) terminal of the 5 volt power supply, and the red alligator clip to the positive terminal. Note the deflection of the display line. Adjust the "Volts/Div" switch to get the most accurate value of this voltage as read from the calibrated graticule.

> **NOTE:** It is essential that the black lead of the oscilloscope be connected to the black (negative) terminal of the power supply, otherwise, a short circuit could occur. The negative terminal of the power supply is connected to its metal box.

2. <u>Time Base</u>

When the "Sec/Div" switch is in a position other than "XY", the electron beam sweeps across the tube face from left to right at a speed indicated by the controls. At the end of its sweep, the beam snaps back quickly to the left hand edge to start over again. In this oscilloscope, the sweep speed is given in 22 calibrated rates ranging from .5 sec/cm to 0.5 μ sec/cm. In addition each of these settings can be made variable and/or made 10 times faster. Gently turn the red variable control knob back and forth, to observe how it feels. Note that it clicks in the extreme clockwise position. In this clockwise "clicked" position, the sweep speed is "Calibrated" and should always be used in this position for time measurements. In addition the red knob can be pulled out to provide a sweep speed ten times faster than otherwise.

You may wish to check the sweep time with a wrist watch for some of the slower sweep spreads.

3. <u>Frequency or Time Measurements</u>

See that the time base is on "Calibrated". Apply the output of the audio generator to the + red and − black vertical inputs of the oscilloscope. Use the sine wave output of the generator and a fairly low frequency. Now turn the SEC/DIV switch to other positions (not "XY") and observe the signal. As the signal is swept across the tube face the vertical oscillations are spread out in time, so that a two dimensional plot of voltage against time is displayed on the screen. It is clear then, that whatever the voltage-time relation $v = v(t)$, the oscilloscope may be used to display it. The signal generator gives two possible types of voltage time variation, the familiar sine wave and the square wave. Examine a square wave output on the oscilloscope and discuss with your partner the periodic rise and fall of voltage that it represents. In later applications of the oscilloscope other wave forms will be encountered.

We will now use the oscilloscope to measure frequencies and periods of vibration. At all times select a calibrated sweep rate that will facilitate easy counting of the undulations. Choose a sine wave form of frequency that is not too low. Line up the left hand maximum of the wave with the first vertical line of the tube face. Divide the number of complete cycles on the tube face by the corresponding number of divisions to get a number N cycles/div. Divide this number by the setting of the sweep rate n sec/div. This gives the frequency N/n cycles/sec (Hz). Compare this with the reading on the audio generator. Repeat for several widely different frequencies.

Sometimes it is desirable in an experiment to choose certain convenient audio generator frequencies for use. Having decided on a frequency, display it on the oscilloscope and see if your partner(s) can figure out what frequency it is. Then let the others in the group do the same. A frequency of 60 Hz should also be examined since this is the frequency of oscillation of the line voltage.

4. The Oscilloscope as an x-y Plotter

The horizontal deflection of the electron beam can be determined by an external voltage. Connect a second test lead to the "Ch 2" input, and connect the two alligator clips together. Adjust the controls for Ch 2 as:

1. Invert button: out
2. Volts/Div: 5 at the 1X side
3. AC-Gnd-DC: DC

Now turn the "SEC/DIV" switch to the counterclockwise "XY" position, and turn the <u>intensity down</u> as much as possible before the dot display disappears. Connect the alligator clips of this Ch 2 test lead to the 5 volt power supply as was done for Ch 1 earlier and note the deflection. Again, measure the voltage as indicated by the oscilloscope.

5. Two Simple Harmonic Vibrations at Right Angles: frequency comparison

When two simple harmonic vibrations at right angles are compounded together, the resulting pattern is called a Lissajous figure for the person who first investigated it. The oscilloscope will display these curves if two appropriate sinusoidal signals are fed simultaneously into the vertical and horizontal inputs. If the frequency ratio f_x/f_y is a simple number (such as 1:1, 1:2, 3:4, etc.) then a pattern similar to Fig. 3 is obtained.

The ratio of the two frequencies may be obtained by counting the number of times the pattern crosses the axes. That is, f_x/f_y = (no. of crossings of y-axis)/(no. of crossings of x-axis). See fig. 3.

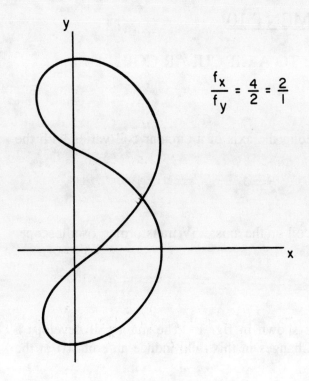

Fig. 3. Lissajous figure

i) <u>Identical Inputs</u>. Take the test leads from both Ch. 1 and 2. Connect the black alligator clips from each to the lower (ground) terminal of the signal generator. Connect the red alligator clips from both Ch. 1 and 2 to the top terminal of the signal generator. Turn up the amplitude on the signal generator and observe the display. This method of analysis will be used in a subsequent lab exercise to study two simple harmonic vibrations of the same frequency.

ii) <u>General case</u>. To study the general case, we will use the 1000 Hz frequency source for one input, and the variable frequency audio oscillator as the other. Remove the test lead alligator clips for Ch 1 (x) from the signal generator. Be sure the alligator clips for Ch 2 (y) remain connected. Connect these alligator clips for Ch 1(x) to the output of the 1000 Hz oscillator. Be sure the black lead is connected to the ground terminal oscillator. Vary the frequency of the audio oscillator to give simple frequency ratios. Careful adjustment of the oscillator frequency is required to give a stable pattern on your screen.

DISCUSSION

Review what you have done and try to understand clearly the four basic uses of the oscilloscope. Discuss

1) Its use as a voltmeter;
2) Its use as an x-y plotter giving the plot of say, voltage against current or one voltage against another;
3) Its use as an x-y plotter giving a plot of voltage against time;
4) Its use as a means of measuring frequency (or period)
   i) using the calibrated time base, and
   ii) using a standard frequency source and Lissajous figures.

EXPERIMENT 10

MAGNETIC FIELD DUE TO A CIRCULAR COIL

OBJECTIVE:

To examine how the magnetic field along the axis of a circular coil varies with the distance from the coil.

EQUIPMENT:

Circular field coil with a small search coil on the axis; 25V transformer; oscilloscope; 23.5Ω resistor; fused switch.

INTRODUCTION:

The circuit used in this experiment is shown in fig. 1. The main coil develops a magnetic field whose axial value is B(x,t). Changes in this field induce an emf $\mathcal{E}(t)$ in the search coil.

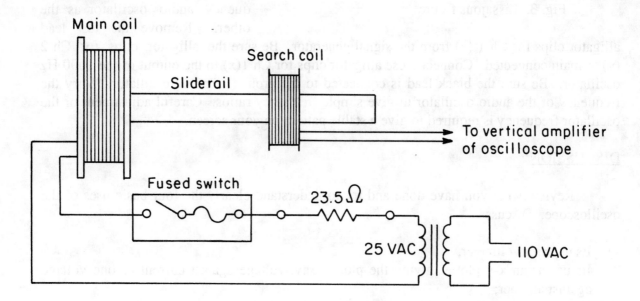

Fig.1. Circuit to examine the magnetic field along the axis of a coil.

The magnetic field B(x,t) along the axis of the main coil of N turns and mean radius a is given by

$$B(x) = \frac{\mu_0 i N a^2}{2(x^2 + a^2)^{3/2}} \tag{1}$$

where i(t) is the current and x is the distance from the center of the coil. Use of the mean radius is a good approximation, provided that the difference between the outer and inner radii is small compared with the mean radius.

To detect the magnetic field of the main coil, the search coil of N_S turns and radius b, is placed at a distance x from the center of the main coil. The flux through the search coil is given by

$$\Phi = \int B(x) \, dA = \pi b^2 B(x) \tag{2}$$

where **B** is considered to be uniform within the search coil, and b is the average of the inner and outer radii of the search coil. If the current i(t) in the main coil varies sinusoidally with amplitude I_0 and circular frequency ω, then eq.(1) can be expressed as

$$B(x) = \frac{\mu_0 N a^2 I_0 \cos(\omega t)}{2(x^2 + a^2)^{3/2}} \tag{3}$$

In this experiment $\omega = 2\pi f$ with $f = 60$Hz, the line frequency.

From Faraday's law of induction, the emf $\mathcal{E}(t)$ induced across the terminals of the search coil is proportional to the rate of change of flux in the search coil and the number of turns, N_S. Thus, with eq. (2),

$$\mathcal{E}(x) = -N_S \frac{d\Phi}{dt} = -\pi b^2 N_S \frac{dB(x)}{dt} \tag{4}$$

The rate of change of the magnetic field is derived from eq. (3),

$$\frac{dB(x)}{dt} = -\frac{\omega \mu_0 N a^2 I_0 \sin(\omega t)}{2(x^2 + a^2)^{3/2}} \tag{5}$$

Substituting eq. (6) into eq. (5) we have an expression for the emf,

$$\mathcal{E}(x) = \frac{N_S \pi b^2 \omega \mu_0 N a^2 I_0 \sin(\omega t)}{2(x^2 + a^2)^{3/2}} \tag{6}$$

This result shows that $\mathcal{E}(t)$ varies sinusoidally with the same frequency as the current i(t) but differs in phase by $\pi/2$.

If just the amplitudes of $\mathcal{E}(x)$ and $B(x)$ are considered, then from eqs. (3) and (6) we find

$$\mathcal{E}_0(x) = N_S \pi b^2 \omega B_0(x) \tag{7}$$

Rearranging eq. (7) for the purpose of this experiment the magnetic field strength can be experimentally determined with use of the equation

$$B_0(x) = \frac{\mathcal{E}_0(x)}{N_s \pi b^2 \omega} \quad . \tag{8}$$

In this experiment you will use the oscilloscope to measure $\mathcal{E}_0$ and subsequently calculate the magnetic field strength. You will then compare the functional form of $B_0(x)$, as experimentally determined, with the field amplitude given by eq. (3).

PREPARATION:

Use eq. (3) to determine the field amplitude ratio $B_0(x)/B_0(0)$ for values of x from 0 to 25cm. The mean radius a is 12.5cm. Plot $B_0(x)/B_0(0)$ vs. x on linear graph paper.

PROCEDURE:

1. Set up the circuit shown in fig. 1. Close the switch to make measurements, and open it at other times to avoid excessive heating of the coil and transformer.

2. Connect the oscilloscope across the 23.5Ω resistor and record the maximum potential. Use Ohm's Law to determine the current amplitude I_0.

3. Connect the oscilloscope to the search coil.

4. Beginning at the center of the main coil, measure the induced emf at 2.5 cm intervals along the slide rail.

ANALYSIS:

1. For each value of $\mathcal{E}_0(x)$, calculate $\mathcal{E}_0(x)/\mathcal{E}_0(0)$ where $\mathcal{E}_0(0)$ is the emf amplitude at x=0.

2. On the graph of $B_0(x)/B_0(0)$, plot $\mathcal{E}_0(x)/\mathcal{E}_0(0)$ vs. x. According to eq. (7), $\mathcal{E}_0(x)/\mathcal{E}_0(0)$ has the same functional form as $B_0(x)/B_0(0)$. Are the plots in agreement?

3. Use eq. (3) to calculate the theoretical values of the magnetic field strength $B_0(x)$ at x=0, 15, and 25cm.

4. Use eq. (8) to calculate the experimental values of the magnetic field amplitude at x=0, 15, and 25cm. Compare these to the theoretical values.

5. How is this coil arrangement like a transformer?

EXPERIMENT 11

RELAXATION OSCILLATOR

OBJECTIVE:

To measure the firing voltage and the extinction voltage of a voltage regulator (VR) tube using a DMM, and to measure the charge time and discharge time of a relaxation oscillator using an oscilloscope.

EQUIPMENT:

300 volt DC power supply, potentiometer voltage divider, board with resistors, capacitor and VR tube, DMM, oscilloscope.

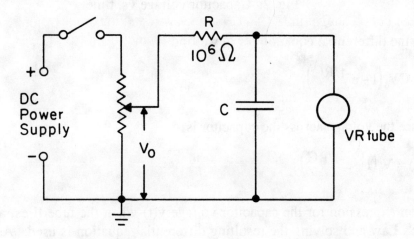

Fig. 1. Relaxation oscillator circuit

The relaxation oscillator used in this experiment consists of the circuit shown in Fig. 1. The VR (voltage regulator) tube is filled with gas which ionizes when the potential difference across it is about 120 V and ceases to conduct at about 110 V. When a voltage V_0 is applied to the circuit, the capacitor charges to a potential difference of V_f, the firing (or ionization) voltage of the VR tube. At this point, the gas in the tube ionizes, the tube lights up, and the capacitor discharges through the gas. When the potential difference across the tube reaches V_e, the extinction voltage, the tube ceases to conduct, the capacitor begins to recharge, and the process is repeated as illustrated in fig. 2.

While the capacitor is charging, the circuit equation as derived from Ohm's Law is

$$V_0 = iR + \frac{q}{c} = \frac{dq}{dt}R + \frac{q}{c} .$$

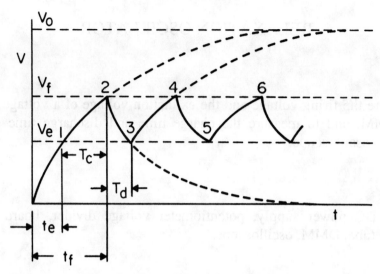

Fig. 2. Capacitor voltage vs. time.

By solving the differential equation for q, we find

$$q = CV_0\left(1-e^{-t/RC}\right)$$

and therefore the voltage across the capacitor is

$$v(t) = V_0\left(1-e^{-t/RC}\right). \tag{1}$$

To obtain an expression for the capacitor voltage v(t) when the tube fires, a similar method using Ohm's Law and solving the resulting differential equation is used. Assuming that the resistance of the tube R_g is much less than R in fig. 1, we determine that

$$\frac{q}{c} = iR_g = -\frac{dq}{dt}R_g$$

which has the solution

$$q = CV_f e^{-t/R_g C}$$

where V_f is the potential across the capacitor when the tube first fires. Then the voltage is

$$v(t) = V_f e^{-t/R_g C} \tag{2}$$

As can be seen from eqs. (1) and (2), the voltage varies with time as illustrated in the charge and discharge portions of fig. 2.

From eq.(1) we see that the charging time T_c will depend on the product RC and the voltages V_0, V_f and V_e. Suppose a charging cycle begins at time t_e when $V=V_e$. Then from eq.(1)

$$V_e = V_0\left(1 - e^{-t_e/RC}\right).$$

At a later time t_f, the capacitor discharges, and $V=V_f$. From eq. (1).

$$V_f = V_0\left(1 - e^{-t_f/RC}\right).$$

The charging time can then be found from

$$T_C = t_f - t_e = RC\,\ln\left[\frac{V_0 - V_e}{V_0 - V_f}\right]. \tag{3}$$

As seen from eq. (3), the charge time T_C varies directly with R and C. In fig. 2 if T_C decreases, the amount of curvature in the voltage waves during charging decreases and the waveform becomes more linear in appearance. Ideally, the discharge time T_d for the capacitor is zero, because the ionized gas in the tube should have no appreciable resistance. If this were true, fig. 2 would show a vertical line while the capacitor is discharging. The oscilloscope, however, shows a small but measurable discharge time, suggesting that the tube contains a non-zero effective resistance while it is conducting.

In this experiment you will first measure the firing voltage V_f and the extinction voltage V_e of the tube using a DMM. Using the oscilloscope, you will then measure charge times T_C, the discharge time T_d and amplitudes of the waveform, $V_f - V_e$.

PROCEDURE:

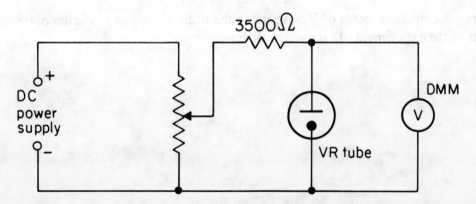

Fig. 3: Circuit to measure V_f and V_e.

1. Set up the circuit shown in fig. 3. The DMM is used to measure the voltage across the tube, and the potentiometer is used to provide a variable voltage to the tube.

2. Prepare a data table to record the measured values of V_f and V_e for several trials.

3. Increase the voltage until the tube fires. The DMM reading will drop slightly at this point. Record V_f, the voltage **just before** the tube fires.

4. Decrease the voltage until the tube ceases to conduct. The DMM will show a slight increase at this point. Record V_e, the voltage **just before** this increase.

5. Repeat steps (3) and (4) several times. Calculate the average value of V_f and V_e, and the uncertainty.

6. Set up the circuit as shown in fig. 1. Connect the oscilloscope across the VR tube. The oscilloscope channel should have its "AC - GND - DC" switch on AC. Use initial values of $R=10^6$ Ω and $C=1$ μF. Set V_0 to at least 200 V.

7. Measure the period, charge time, and discharge time using the division scale and the number of divisions shown on the oscilloscope. In addition, record the amplitude of the wave.

8. Repeat step (7) using a higher value of V_0.

9. Replace the capacitor with a smaller one and repeat step (7). It may be necessary to adjust V_0 to obtain oscillation.

ANALYSIS:

1. Using eq. (3), calculate the expected charge times and compare them to the measured values.

2. Compare your measurements of $V_f - V_e$ using the oscilloscope to the value obtained in the first part of the experiment. Discuss any differences.

EXPERIMENT 12

CURRENT BALANCE

OBJECTIVE:

To measure the permeability of free space, μ_0 by measuring the force between two parallel wires which carry a current I.

EQUIPMENT:

Current balance, 6V DC power supply, ammeter, adjustable resistor, set of fractional masses, telescope with stand and scale, meter stick.

INTRODUCTION:

The magnetic field at a distance d from an infinitely long wire carrying a current I is given by

$$B = \frac{\mu_0 I}{2\pi d} \tag{1}$$

and is perpendicular to the wire. The force on a wire carrying current I and having (vector) length **L** is

$$\mathbf{F} = I\,\mathbf{L}\times\mathbf{B} \,. \tag{2}$$

If the current is perpendicular to the magnetic field, then

$$F = ILB \,. \tag{3}$$

If two parallel wires carry a current I and are a distance d apart, then the force between them is

$$F = \frac{\mu_0 I^2 L}{2\pi d} \tag{4}$$

where L is the length of one wire and the other is infinitely long.

As can be seen from the form of eq. (4), the slope of a F vs. I^2 graph will yield slope ($\mu_0 L/\pi d$). From the slope μ_0, the permeability of free space can be determined.

In this experiment you will set up the current balance shown in fig. 1. It consists of a baseboard which supports a long stationary conductor. A second conductor is balanced just above the first one and is supported by knife-edges. A current is passed in opposite directions in the two parallel conductors. This causes a repulsive force and the upper conductor rises. A weight is placed in a pan on the upper conductor, thus causing the conductor to descend to its original position. The position of the conductor is observed by means of a mirror mounted on the beam between the knife-edges.

Fig. 1. The current balance.

To obtain μ_0 from the slope of the F vs. I^2 graph, one needs to know the distance between the centers of the conductors. See fig. 2. It can be shown that

$$2\frac{d_0}{a} = \frac{D}{b} \qquad (5)$$

where D is the difference in the scale readings when the conductor is displaced from equilibrium and the conductors make contact, and b, a, and d_0 are defined in fig. 2. Since $d = d_0 + 2R$, the radius of the conductor, eq. (5) can be rewritten as

$$d = \left(\frac{Da}{2b} + 2R\right) . \qquad (6)$$

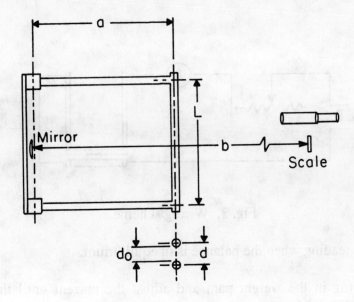

Fig. 2. Schematic top view of moveable current arm.

PROCEDURE:

1. Set up the current balance. First, use the lift mechanism to accurately position the balance frame on the bearing blocks. Second, adjust the counterweight behind the mirror until the frame oscillates freely. When it comes to rest, the conductor should be a few millimeters above the stationary conductor. Third, adjust the counterweight below the mirror until the period of oscillation of the frame is 1 to 2 seconds.

2. Align the two conductors by placing a coin on the weight pan and bringing the conductors into contact.

3. Set up the telescope and scale on the stand so that the telescope is focused on the reflection of the scale in the mirror. The scale should be 1 to 1.5 meters from the mirror. An adjustment screw behind the mirror permits the tilt of the mirror to be varied. The tilt of the balance frame is indicated by the reading of the scale through the telescope.

4. Connect the current balance to the power source as shown schematically in fig. 3. The wires connected to the binding posts should be at right angles to the conductors, because the magnetic fields of these wires could provide unwanted forces on the frame.

5. Referring once again to fig. 2, measure L, the length of the upper conductor. This is the distance from the center of one supporting bar to the center of the other. Measure *a*, the distance from the knife-edge to the center of the front bar. Measure *a* on the other side of the balance frame, and average the two values.

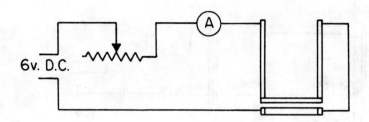

Fig. 3. Wiring scheme.

6. Record the scale reading when the balance is in equilibrium.

7. Place about 20 mg in the weight pan, and adjust the current until the conductors are balanced. Record the current.

8. Increase the total mass to about 30 mg. Adjust the current until the conductors are balanced, and record the current.

9. Continue adding mass up to 100 mg or until you have sufficient data. Record the current required to balance the conductors for each mass added. It is a good procedure to record all readings without lifting the balance frame with the lift mechanism. This is possible if weights are added and removed carefully with tweezers to avoid bumping the balance frame.

ANALYSIS:

1. Calculate the gravitational force F (weight in newtons) for each mass, and the square of the current, I^2.

2. Make a plot of F vs. I^2 on linear paper. As an alternative, you may plot F vs. I on log-log paper, then you can determine the I-squared dependence.

3. Calculate μ_0 from the slope (if you plotted F vs. I^2), and compare it to the expected value of $4\pi \times 10^{-7}$ kg m/C^2.

4. Estimate the statistical uncertainty of the slope, and use it with the uncertainty in d and L to find the uncertainty in your experimental value of μ_0.

EXPERIMENT 13

ELEMENTS OF A. C. CIRCUITS

OBJECTIVE:

To determine the voltages, reactances, and relative phases of the components of an LRC circuit using direct measurement, and a phasor diagram.

EQUIPMENT:

Circuit board with inductor, 1000Ω resistor, and a capacitor; low-voltage transformer; DMM; oscilloscope; compass; protractor.

INTRODUCTION:

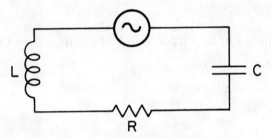

Fig.1: Series LRC circuit

Consider a sinusoidal signal such that the voltage as a function of time is given by

$$v = V \sin(\omega t) \tag{1}$$

where v is the instantaneous voltage, V is the maximum voltage amplitude of the signal, and ω (=$2\pi f$) is the circular frequency. In the series LRC circuit of fig. 1, the circuit components have respective voltages of v_L, v_R, and v_C which also vary sinusoidally, but not with the same phase. As shown in fig. 2(a), the voltage across the resistor v_R is **always** in phase with the current. The voltage across the inductor v_L leads the current by $\pi/2$, and the voltage across the capacitor v_C lags the current by $\pi/2$.

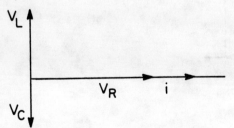

Fig. 2(a). Voltage diagram

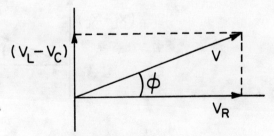

Fig. 2(b). Resultant voltage and phase

The relative phases (angles) and maximum voltages (magnitudes) of the circuit components can be represented by vectors (typically called phasors) as in fig. 2(b). The current phasor I is conventionally drawn horizontally to the right. The magnitude of the applied voltage V and its phase φ is represented in the phasor diagram by the resultant vector. The current i is given by

$$i = I \sin(\omega t) \ . \tag{2}$$

The applied voltage leads the current by phase angle φ and is expressed as

$$v = V \sin(\omega t + \varphi) \ . \tag{3}$$

From fig. 2(b) and simple geometry, the phase constant φ is given by

$$\tan \varphi = \frac{V_L - V_C}{V_R} \tag{4}$$

The current amplitude I in the circuit can be determined by Ohm's Law for AC,

$$I = \frac{V}{Z} \tag{5}$$

where Z, the impedance, is defined as the combination of resistances and reactances

$$Z = \sqrt{R^2 + (X_L - X_C)^2} \ . \tag{6}$$

The mathematical form of eq. (6) suggests that an impedance diagram can be drawn as in fig. 3. The resistance R is drawn horizontally to the right; the inductive reactance, given by $X_L = \omega L$, is drawn vertically upward; and the capacitive reactance, given by $X_C = 1/\omega C$, is then subtracted from X_L. The impedance Z is the vector sum of the reactances and the resistances.

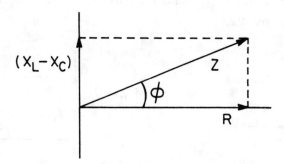

Fig. 3: Impedance Diagram

For a series circuit, the current amplitude is equal through each circuit element. Therefore, Ohm's Law can be expressed for each circuit component in the form

$$I = \frac{V_R}{R} , \qquad (7)$$

$$I = \frac{V_L}{X_L} , \qquad (8)$$

$$I = \frac{V_C}{X_C} . \qquad (9)$$

The AC ammeter (like the voltmeter) does not measure instantaneous current *i*, but acts like a simple integrator, determining the average of the square of the current over one cycle. The square root of this average is called the root mean square current, or more commonly, the "rms current." The maximum current I is related to the rms current according to the equation

$$I_{rms} = \frac{I}{\sqrt{2}} . \qquad (10)$$

Likewise, for rms voltages. The current-voltage relations hold for rms values, just as they do for amplitudes.

In actual circuits the inductor will also contain an intrinsic resistance r, which will have to be accounted for. Hence, the actual voltage across the inductor is

$$V_{Lr} = I\sqrt{X_L^2 + r^2} . \qquad (11)$$

The phasor representing V_{Lr} can be thought of as consisting principally of the purely inductive voltage component plus a small resistive component as shown in fig. 4.

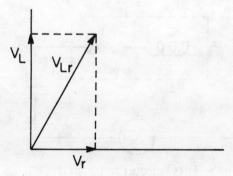

Fig. 4. The phasor V_{Lr}

In all expressions which follow, the effects of the intrinsic resistance r will be included.

In this experiment you will use a DMM to determine r (a D.C. measurement) and the currents and voltages across the components of LR and RC circuits (A.C.). You will then predict the voltage-current relationships in an RLC circuit and use the DMM to verify these relationships. Finally, you will use an oscilloscope to check the applied voltage in the RLC circuit and determine the phase relation between V and I. (Note: Always determine I through measurement of V_R, using $I = V_R/R$.)

PROCEDURE:

1. Create a data table similar to the one below to record data and calculations.

Circuit	V	V_R	V_{Lr}	V_C	V_{LR}	$I=V_R/R$	X_C	X_L	C	L	r	φ
D.C. Measurement												
LR (DMM)												
RC (DMM)												
LRC (Predicted)	(Graph)	(DMM)										(Graph)
RLC (DMM)												
RLC (Oscilloscope)												

Table 1. Circuit parameters

2. Set up the LRC circuit in fig. 5, and record the nominal values L, R, and C. The inductor is not ideal, so it contains a resistance represented by r.

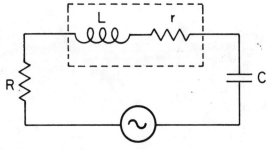

Fig. 5. LRC circuit to be analyzed.

3. Using the DMM, measure the resistance r (a D.C. measurement).

4. With a wire, short out the capacitor to form an LR circuit. Check that you did this properly. Measure the voltages relevant to this circuit.

5. Short out the inductor to create an RC circuit. Repeat Step (4) for the relevant voltages.

6. For the complete LRC circuit measure I and, using the DMM on the correct voltmeter range, measure the voltages V, V_R, V_{Lr}, V_C, and V_{LR}, the voltage across the inductor and the resistor. (All A.C. measurements with the DMM are **rms** values.)

7. Using the oscilloscope, measure the peak-to-peak voltage of the source. Compare this to the rms value obtained previously by using eq. (10) modified for voltage, and $V_{pp}=2V$.

8. Determine the phase constant φ. Since the phase is measured relative to the resistor, place the oscilloscope probe across the resistor and note which end is ground. Adjust the SEC/DIV so that one cycle occupies 8 major divisions on the oscilloscope screen. Since one cycle corresponds to an angular measurement of 360°, each division is 45°. Now place the probe across the source, keeping the ground in the same **relative** position as before. Record the number of divisions that the signal has shifted, and calculate φ in degrees.

ANALYSIS:

1. Use eq. (7) to calculate the current through the resistor from V_R in all cases. This is also the current throughout the circuit.

2. Use eq. (11) to calculate X_L from your measurements on the LR circuit, and eq. (9) to calculate X_C from measurements on the RC circuit.

3. Using $f=60Hz$, calculate ω. Determine L and C and compare to the given nominal values.

4. Draw the phasor diagram as in fig. 2(b), being sure to include V_r with V_R (note $V_r=Ir=V_R(r/R)$, and using as the inductive voltage only IX_L, which excludes the effect of the intrinsic resistance r (note $IX_L = \sqrt{V_{Lr}^2 - (Ir)^2}$). Using the diagram, determine V and φ.

5. Calculate the impedance Z, correctly accounting for r, R, X_L, X_C. Compare this with the measured value obtained through the equation Z=V/I, or simply $(V/V_R)R$.

EXPERIMENT 14

A.C. SERIES CIRCUITS: RESONANCE AND FORCED OSCILLATION

OBJECTIVE:

To further analyze the electrical properties of an LRC circuit.

EQUIPMENT:

Frequency generator; oscilloscope; LRC series circuit board; switch.

INTRODUCTION:

In Experiment 13 you kept the source voltage, the waveshape (amplitudes, etc.), and the frequency constant and studied the relative phases and voltages across the circuit components. In this experiment you will vary the source frequency and study its effect on current.

Assume that the applied AC signal has a time dependence according to

$$v = V \sin(\omega t) \, . \tag{1}$$

The magnitude and sign of v at any instant are determined by the values of the parameters t, ω, and V. The source voltage oscillates sinusoidally with circular frequency $\omega = 2\pi f$ and has a constant amplitude V.

As in a DC circuit, the current is determined by the applied voltage and the components in the circuit. As explained in Experiment 13 the current is given by

$$i = I \sin(\omega t - \varphi) \, . \tag{2}$$

The current oscillates with the same frequency as the applied voltage, but is out of phase by φ. This says that the current lags the applied voltage by a phase constant φ. Eqs. (1) and (2) may also be written as in Experiment 13, where the voltage leads the current by a phase constant φ. The current amplitude I is determined according to Ohm's Law for AC,

$$I = \frac{V}{Z} \, . \tag{3}$$

The impedance Z is determined by the resistance R and the reactances X_L and X_C according to the relation

$$Z = \sqrt{R^2 + (X_L - X_C)^2} \quad , \tag{4}$$

where

$$X_L = \omega L \tag{5}$$

$$X_C = \frac{1}{\omega C} \quad . \tag{6}$$

As seen from an impedence diagram (Experiment 13), the phase angle φ is given through

$$\tan \varphi = \frac{(X_L - X_C)}{R} \quad . \tag{7}$$

If the amplitude V of the source voltage is held constant and the frequency ω is varied, the current amplitude I varies inversely with the impedance Z. Therefore, I will be largest for the smallest possible value of Z. From eq. (4), the smallest value of Z occurs when

$$X_L = X_C \quad , \tag{8}$$

and therefore

$$\omega L = \frac{1}{\omega C} \quad . \tag{9}$$

Solving this equation for ω gives the resonant frequency

$$\omega_0 = \frac{1}{\sqrt{LC}} \quad . \tag{10}$$

It is this particular frequency that yields the largest possible current amplitude. As seen in eq. (10), ω_0 does not depend on the resistance R in the circuit, although the current amplitude I indeed does depend upon R. This can be seen in fig.1 where the current amplitude is plotted vs. the frequency, with V, C, and L being fixed and R being set differently for each curve.

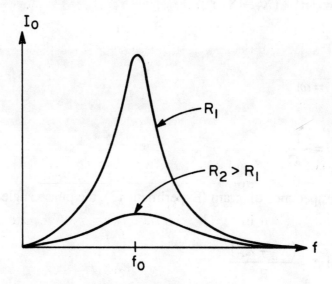

Fig.1. Current amplitude I vs. frequency ω for various values of R.

If $X_L=X_C$, then from eq. (6) $\varphi=0$. So the resonant frequency can also be determined from a plot of φ vs. ω as in fig. 2.

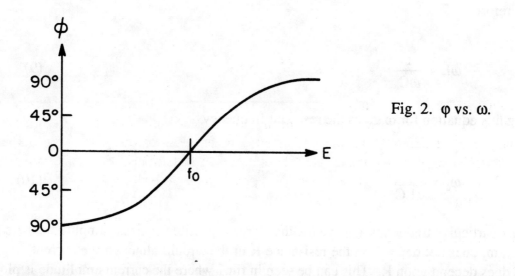

Fig. 2. φ vs. ω.

In this experiment you will make measurements of voltage V_R and frequency f using an oscilloscope. For any frequency, the current can be calculated from V_R using Ohm's Law; thus,

$$I = \frac{V_R}{R} = \frac{V_{R_{PP}}}{2R} \quad , \tag{11}$$

where $V_{R_{PP}}$ is measured peak to peak. Knowing I and f, you can make graphs similar to those of fig. 1 to find the resonant frequency. You can compare this measured value of f_0 to the expected value obtained from eq. (10). Remember that the oscilloscope is used to measure f, and $\omega = 2\pi f$.

PROCEDURE:

1. Familiarize yourself with the frequency generator. Connect the oscilloscope across the output of the generator. Observe the effects of the voltage amplitude selector and the frequency dial on the waveform.

2. Set up the circuit shown in fig. 3 with the switch closed, and record the values of L, R, and C.

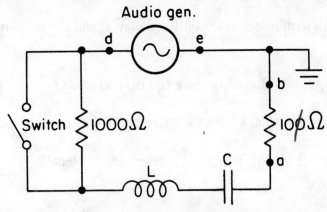

Fig. 3. Series LRC circuit

3. The current can be calculated using the voltage across the 100Ω resistor V_{100}, and eq.(11). To measure V_{100}, connect Channel 1 of the oscilloscope across the resistor with the ground (black lead) at point b. Connect Channel 2 of the oscilloscope across the output of the frequency generator with the ground at point e, the low terminal. So now Channel 1 should read V_{100}, and Channel 2 should read V. You can observe both signals on the oscillocope at the same time by setting the "Vertical Mode" switch to "both."

4. Set the frequency at about 3 KHz. Maintain the output voltage at 8 volts peak to peak. You can do this by adjusting the AMPLITUDE on the generator.

5. Measure the peak to peak voltage across the 100Ω resistor.

6. Measure the frequency. To do this with the oscillocope, count the number of cycles that occurs over a certain number of horizontal scale divisions. The frequency is then given by

$$f = \frac{\# \text{ cycles}}{\# \text{ div}} \times \frac{\# \text{ div}}{\sec} \ . \tag{12}$$

7. Resetting the frequency at about 1 KHz intervals in the 3 to 10 KHz range, measure the corresponding voltages and frequencies. Take additional measurements near the peak of the I vs. f curve, so that the curve is well-defined. Also be sure to determine the frequency and current at the very peak of the curve. It is a good idea to make a rough sketch of the curve as you take the data, so you can avoid any careless errors.

8. Open the switch and repeat steps (3) - (7) for R=100 +1000=1100Ω

ANALYSIS:

1. For each frequency setting, determine the current I and the frequency f.

2. Plot a graph of I vs. f for R=100Ω. On the same graph paper, plot I vs. f for R=1100Ω.

3. From the data, determine the resonant frequency f_0 and the maximum current $I(f_0)$ for both plots.

4. Calculate the expected value of f_0 using eq.(10) and $\omega=2\pi f$.

5. Compare your experimental values for the resonant frequency to the theoretical value.

6. From your values of $I(f_0)_{1100}$ and $I(f_0)_{100}$, determine r. Recall,

$$\frac{I(f_0)_{1100}}{I(f_0)_{100}} = \frac{100+r}{1100+r}$$

so that

$$r\left(1 - \frac{I(f_0)_{1100}}{I(f_0)_{100}}\right) = \left(1110 \frac{I(f_0)_{1100}}{I(f_0)_{100}} - 100\right)$$

or

$$r = \frac{\left(1100 \frac{I(f_0)_{1100}}{I(f_0)_{100}}\right)}{\left(1 - \frac{I(f_0)_{1100}}{I(f_0)_{100}}\right)} = \frac{(1100\, I(f_0)_{1100} - 100\, I(f_0)_{100})}{I(f_0)_{100} - I(f_0)_{1100}}.$$

58

EXPERIMENT 15

COMPUTERIZED PENDULUM

OBJECTIVE:

    A. To study the motion of a real pendulum
    B. To make a measurement of the acceleration of gravity
    C. To study the dependence of period on amplitude for a pendulum.
    D. To observe the damping due to friction for a pendulum.

EQUIPMENT:

Long thin rod attached to shaft of a potentiometer (variable resistor), adjustable mass, meter stick, computer with clock and analog to digital converter.

INTRODUCTION:

One example of motion that is nearly simple harmonic is the swing of a simple pendulum. This consists of a mass suspended at the end of a long light rod. The equilibrium position for this apparatus is for the rod to be vertical. If displaced from this position, there are two forces acting on the bob. One is the weight of the bob, which acts vertically downward. The other is the tension in the rod which acts along the rod. The resultant of these two forces (vector sum) is a force tending to pull the bob back to its equilibrium position. If the angle of displacement of the rod from the vertical is small, the motion of the bob is approximately simple harmonic, with the period given by

$$T = 2\pi \sqrt{\frac{\ell}{g}} \tag{1}$$

where T is the period, ℓ is the length of the pendulum, and g is the acceleration due to gravity. If the angle of displacement is not small, the period also depends on the angle of maximum displacement.

The pendulum rod is attached to a bearing which has some friction. This will tend to slow the pendulum and eventually cause it to stop. The decrease of amplitude with time is approximately described by a decaying exponential.

DISCUSSION:

The potentiometer contains a fixed resistance element and a sliding contact attached to a shaft which can rotate. As the shaft is rotated, the contact slides along the resistance. This device is used to electrically measure the angular position of the shaft. A potential of ten volts is maintained between opposite ends of the resistance by power supplies inside the computer. Thus, the electrical potential of the sliding contact is a measure of angle. This potential is converted into an eight bit binary number (0 to 255) and read by the computer. This conversion is done by the analog to digital converter (ADC). As the thin rod with the mass on it swings back and forth, it rotates the shaft, changing the position of the sliding contact on the resistance and its electrical potential. The clock supplies an accurate timing signal 1000 times a second. This timing signal is used by the computer to tell it to measure the voltage from the sliding contact again. Thus this voltage, and hence the angle are measured 1000 times a second.

The computer program has three main parts. The first part is calibration. This is done by having the computer obtain the digitized voltages corresponding to the pendulum rod being vertical and horizontal. For instance, the difference between these two readings corresponds to 90 degrees. This calibrates the apparatus for data accumulation in the last two parts.

The next part is to measure the period for various amplitudes. The computer reads and stores 8000 data points for the position of the pendulum in 8 seconds. It then graphs these points as a function of time, and determines the amplitude and period from a selection of the data. Several such data points are collected in this part. When enough data has been collected to show the dependence of period on amplitude, this part is concluded, and the data is stored on disk.

The last part of the experiment involves starting the pendulum swinging at some large angle slightly less than 90° and observing the amplitude decrease with time. This part collects 8000 data points in 8 seconds as before, finds the amplitude as before, and puts the amplitude in a table. It then goes back to repeat, collecting another 8000 samples. The period is ignored. In this way, the decrease of amplitude with time can be observed.

PROCEDURE:

1. Obtain a diskette from your instructor. Carefully hold the diskette by placing the thumb on the label and the index finger on the opposite side. DO NOT touch the recording surface. Place the diskette in the disk drive, close the door and turn on the computer (switch to left). You will be asked to give the number of people in your group and your names.

2. The screen will display the main menu "Analysis of Pendulum Motion". Calibrate the angles. Be sure the rod is perfectly horizontal when asked to do so. (What is the most accurate way to do this?) If the calibration was not correct, you can do it over by returning from the main menu.

3. Fix the mass on the rod, near the end, but not at the end. Measure the pendulum length. (Between which points do you measure the pendulum length?) From the main menu ("Analysis of Pendulum Motion") go to "Period vs. Amplitude". Take several data points (at least 10) of the period at given amplitudes. Be sure to have several at small angles and

some near 90°. Do not exceed 90° for amplitude. Observe the data as a graph or table. When enough data is collected, select "Store & Back to Menu", to store these data on disk and return to the main menu.

4. From the main menu go to "Amplitude vs. Time". Start the pendulum swinging with an amplitude slightly less than 90°. When the data is accumulated, observe it as a graph or table. If it is unsatisfactory take another set, and the previous set will be lost. If the data is satisfactory go to "Store & Back to Menu". This will store the data on diskette.

5. If you have completed "Period vs. Amplitude" and "Amplitude vs. Time", select "Finished".

6. Remove the diskette and take it to the printer station for printing and plotting.

ANALYSIS:

1. Estimate the period of the pendulum for very small amplitude, and use this time to calculate the value of the acceleration of gravity.

2. Observe the change in period with amplitude. Explain why it should increase or decrease.

3. Find the time constant of the decay in amplitude of the pendulum. The time constant is the time for the amplitude to decrease by e. Here e has the value 2.718281828 .

QUESTIONS:

1. What does your data for small amplitude look like? Explain its appearance.

2. What is the length of the pendulum, in terms of the elements of the apparatus? From where to where do you measure the "pendulum length"?

3. How did you place the pendulum rod in the horizontal position for calibration? Suggest a more precise method. What was the calibration error for the method you used?

CONCLUSIONS:

1. What was shown about the relationship of the pendulum amplitude to period?
2. What was shown about the damping (friction) of the pendulum?

PHYSICS 204

EXPERIMENTS

EXPERIMENT 1

TRANSVERSE WAVES

OBJECTIVE:

To study the nature of transverse waves in (a) ropes, (b) strings and (c) dielectric media.

REFERENCES:

Physics by Tipler, Fundamentals of Physics by Halliday and Resnick or Vibrations and Waves by French.

EQUIPMENT:

Experiment (a)
Heavy clothes-line, kilogram masses, kilogram hanger, pulley, stop, meter stick and a beam balance.

Experiment (b)
String vibrator, string, pulley, clamps, 50 gram hanger, set of masses and a meter stick.

Experiment (c)
Co-axial cable, oscilloscope, signal generator, two decade resistance boxes, capacitor and hook up wire.

INTRODUCTION:

The speed, v, of a wave depends upon the properties of the medium through which it propagates. In a rope or string under tension, T, and having mass per unit length, μ, the relation

$$v = \left(\frac{T}{\mu}\right)^{1/2} \tag{1}$$

will apply to a high degree of approximation regardless of whether the "wave" is a continuous train of sinusoidal waves or is a pulse of arbitrary shape. Should the wave be purely sinusoidal with difinite wavelength, λ, and frequency, f, then we may also write

$$v = \lambda f . \tag{2}$$

There need not be a medium for light and other electromagnetic waves, which can propagate in a vacuum with speed $c = 2.998 \times 10^8$ m/s. Electromagnetic waves propagating in a dielectric medium travel with a speed which is less than c. The index of refraction, n, of the dielectric material is then defined by

$$n = \frac{c}{v} \tag{3}$$

Maxwell's electromagnetic theory shows that if the dielectric is non-magnetic, the index of refraction may be simply related to the dielectric constant, κ, of the material by

$$n = \sqrt{\kappa} \quad . \tag{4}$$

PROCEDURE:

Experiment (a)

1. Set up the apparatus in fig.1 in order to measure the speed of a pulse as a function of tension for a long rope with fixed ends.

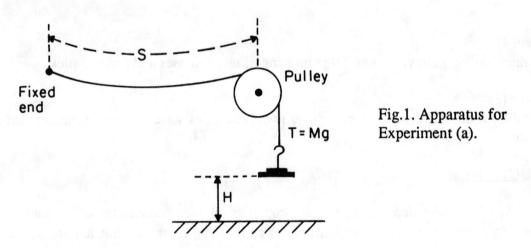

Fig.1. Apparatus for Experiment (a).

2. Measure the mass of the kilogram hanger.

3. Beginning with only the kilogram hanger on the free end of the rope, measure the length L of the rope, the distance A from the ground to the top of the pulley, and the height H of the mass hanger. You can calculate S according to

$$S = L + H - A \quad . \tag{5}$$

4. Create a pulse by sharply plucking the rope, and measure the time it takes for the pulse to traverse the length S ten times. Determine the speed of the wave.

5. Repeat step (4) for 4 others tensions. Be sure to record the total mass M on the rope and the length S, for each new tension.

6. Tabulate the speed of the pulse v and the tension Mg on the rope.

7. Determine μ, the mass per unit length of the rope, by weighing the rope and dividing by its length.

Experiment (b)
1. Set up the apparatus shown in fig.2. The 120 Hz vibrator sends a succession of waves along the string, and under suitable conditions these are reflected at the pulley to produce standing waves. The vibrator should be approximately 1 meter from the pulley.

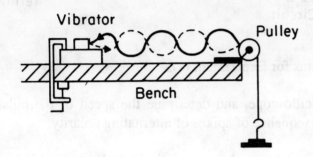

Fig.2. Apparatus for Experiment (b).

2. With the 50g mass hanger in place, slowly add masses until the string vibrates with 5 complete loops. Make fine adjustments by using smaller masses or by sliding the vibrator toward or away from the pulley until maximum amplitude and stationary nodes are obtained. This is known as resonance.

3. Measure the distance between adjacent nodes and determine the wavelength. (The wavelength is twice the distance between nodes.)

4. Record the total mass which provides the tension.

5. Add more mass to the pan so there are 4 standing waves, and repeat the experiment.

6. Repeat with 3 standing waves.

7. Tabulate the tension on the string and the speed of the wave (as determined from eq.(2))

8. Record μ, the mass per unit length of the string, as indicated on the base of the vibrator.

Experiment (c)
1. Set up the apparatus shown in fig.3. This is the electrical analog of Experiment (a). Pulses are generated by means of a square wave generator and an R-C differentiation circuit, travel to the end of the co-axial cable, reflect as dictated by the nature of the termination, and return with reduced amplitude to the beginning.

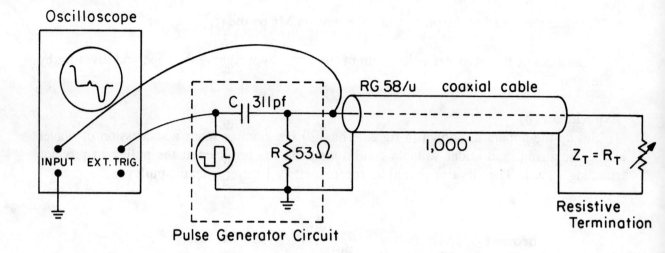

Fig.3. Apparatus for Experiment (c).

2. Measure the transit time using the oscilloscope, and determine the speed of the pulse. The pulses appear on the oscilloscope as a sequence of spikes of alternating polarity.

3. The coaxial cable used in this experiment has a **characteristic impedance** Z_0 of about 53Ω. The reflection coefficient r is defined as the ratio of the voltage amplitudes of the reflected and incident waves. For a lossless cable, r is given by

$$r = \frac{V_r}{V_i} = \frac{Z_T - Z_0}{Z_T + Z_0} \qquad (6)$$

where Z_T is the impedance of the termination or load. This predicts that for $Z_T = Z_0$, the reflection coefficient should be zero, in which case all the incident energy is totally absorbed by the load. The load is then said to be matched to the transmission line. However, if $Z_T = 0$ or $Z_T = \infty$, then r = −1 or +1 respectively; and the incident energy is entirely reflected. A negative r indicates that the reflected pulse is inverted, or equivalently, has undergone a phase change of π upon reflection.

In this experiment, $Z_T = R_T$ in one case, and in the other $R_T = \infty$. Set $R_T = 53Ω$, then use an open circuit, and measure the amplitude and phase of the reflected pulse in each case. Measure the round trip travel time for the reflected pulse.

ANALYSIS:

Experiment (a)

1. Plot v^2 vs. T on linear graph paper. Determine μ from the slope and compare it to the directly measured value.

2. Is μ really constant? If it is not, how are your results affected?

Experiment (b)

1. Plot v^2 vs. T on linear paper, and determine μ. Compare your calculated value with the expected value indicated on the base of the vibrator.

2. If the tension required for resonance in step 6 of Experiment (b) is increased by a factor of 3, will there still be a resonance? What if the tension is increased by a factor of 4?

3. If the vibrator is powered by 60Hz current from the power outlet, why does it vibrate with frequency 120 Hz?

Experiment (c)

1. Use eq.(3) and the measured signal speed (= 2L/ Round trip time) to calculate the index of refraction n.

2. Use eq.(4) to calculate the dielectric constant of the insulating material (probably polyethylene) in the cable.

3. Can you describe how the R-C differentiation circuit in Experiment (c) works?

EXPERIMENT 2

LONGITUDINAL WAVES

OBJECTIVE:

To study the nature of longitudinal sound waves in air and to measure the speed of propagation.

REFERENCES:

Physics by Tipler, Fundamentals of Physics by Halliday and Resnick, Sound by A.J Jones, Vibrations and Waves by French.

EQUIPMENT:

Experiment (a)
Kundt's tube; cork dust; meter stick; thermometer; powdered rosin and cloth.

Experiments (b) and (c)
Ultrasonic transducers; oscillocope; frequency generator; low inductance choke; meter stick.

INTRODUCTION:

The speed of sound in an ideal gas is given by

$$v = \left[\frac{\gamma P}{\rho}\right]^{1/2} \qquad (1)$$

where γ is the ratio specific heat at constant pressure to specific heat at constant volume ($\gamma=1.4$ for air), P is the mean pressure, and ρ is the mass density of the gas. By using the equation of state for an ideal gas, PV=nRT, eq.(1) may be rewritten as

$$v = \left[\frac{\gamma n RT}{V\rho}\right]^{1/2} \qquad (2)$$

where V is the volume occupied by n moles of gas at absolute temperature T and R is the gas constant. Since $V\rho$ is the mass of the gas m and n/m is just the molecular weight M, eq.(2) can be simplified to

$$v = \left[\frac{\gamma RT}{M}\right]^{1/2} \qquad (3)$$

From eq.(3), the ratio of the speeds of sound in the same gas at different temperatures can be solved for the speed at temperature T such that

$$v(T) = v(T_0)\left[\frac{T}{T_0}\right]^{1/2} \qquad (4)$$

For T=273.15K (0°C), the speed of sound in dry air is 331.45 m/s; therefore, according to eq.(4), the speed of sound in air at temperature T is

$$v(T) = 331.45\left[\frac{T}{273.15}\right]^{1/2} \qquad (5)$$

In **Experiment (a)**, you will use Kundt's tube, an instrument designed to measure the speed of sound in gases or solids. The apparatus, shown in fig.1, consists of a metal rod AB clamped at its midpoint C. The usefulness of the rod is that it creates longitudnal vibrations when it is stroked with a rosined cloth. The end of the rod extending into the glass tube is

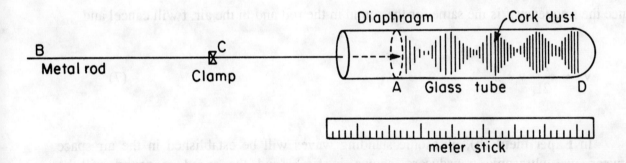

Fig.1. Kundt's tube.

fitted with a rigid circular diaphragm which efficiently transmits sound energy from the rod to the air in the tube. In the tube is dry cork dust which has been spread in a thin uniform layer.

The rod is to be stroked near the free end B so that it emits a loud note of characteristic frequency and sets up sympathetic vibrations in the air in the tube. The tube is then tuned. This is done by varying the length of the air column AD until the cork dust is violently agitated, accumulates and finally settles at the nodes, since standing waves have been produced in the column. At this point the column is in resonance and the distance between the nodes is $\lambda_a/2$ where λ_a is the wavelength of the sound in air. For the rod, the clamp in the middle forces a node at that point while the ends vibrate as antinodes, as shown in fig.2.

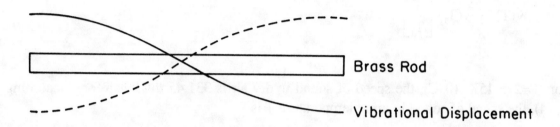

Fig.2. Longitudinal sound amplitude for lowest frequency resonance when the rod is clamped at its midpoint.

As seen in fig.2, the length L of the rod is $\lambda_r/2$ where λ_r is the wavelength of the sound in the rod. The ratio of the speed of sound in the air to that in the rod is

$$\frac{v_a}{v_r} = \frac{f\lambda_a}{f\lambda_r} \qquad (6)$$

Since the frequency f is the same for the sound in the rod and in the air, f will cancel and

$$v_a = \frac{\lambda_a}{2L} v_r \qquad (7)$$

In **Experiment (b)**, ultrasonic standing waves will be established in the air space between two ultrasonic transducers, shown in fig.3, and the speed of sound will be determined. The transducers are made of barium titanate, a polycrystalline ceramic which is permanently electrically polarized. If an AC voltage is then applied, the material will elongate and contract at the same frequency and thus act as an acoustic transmitter. Conversely, if a pressure wave (sound wave) is applied to the device it will generate an alternating voltage of the same frequency and act as a detector. The low inductance choke in figs. 3 and 4 serves as a low pass filter, providing a low impedance path to ground for 60 Hz and other extraneous low frequency noise.

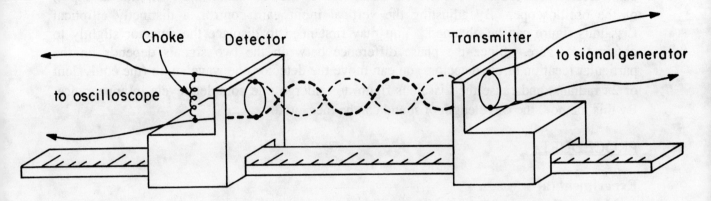

Fig.3. Standing waves produced between two ultrasonic transducers.

In **Experiment (c)**, you will investigate the variation of phase of a wave by means of Lissajous figures. Suppose that two waves of equal frequency and of voltage amplitude A_1 and A_2 are applied respectively to the vertical and horizontal input of an oscilloscope. In other words, they are applied at right angles to each other. The ensuing pattern, which takes the form of an ellipse, is a particular type of Lissajous figure and will depend upon the phase difference δ of the waves. As the phase difference is increased from δ to $\delta + 2\pi$, the Lissajous figure will continuously change its appearance until finally it returns to its original form. The apparatus used to demonstrate this is shown in fig.4. It consists of a detector

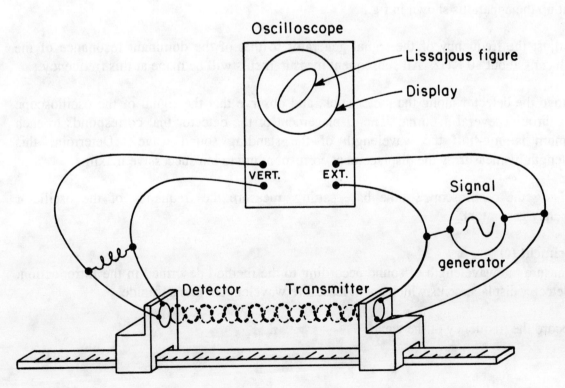

Fig.4. Apparatus to demonstrate Lissajous figures.

connected to the vertical input and the transmitter to the horizontal input (or external input) of the oscilloscope. By adjusting the vertical input gain control, a distinctly elliptical Lissajous figure can be obtained. You may first need to displace the detector slightly to obtain the figure. Since the phase difference between the two signals depends on the particular location of the detector, you can move the detector one wavelength (the equivalent of 2π radians) and cause the Lissajous figure to undergo one complete cycle in appearance. By this method, the wavelength of sound can be measured.

PROCEDURE:

Experiment (a)
1. Measure the temperature of the room T.

2. Set up the apparatus of fig.1.

3. Following the instructions in the Introduction, tune the tube.

4. With a meterstick, measure the distance between the nodes in the tube, and use this to calculate λ_a.

5. Measure the length of the rod L.

Experiment (b)
1. Set up the apparatus shown in fig.3.

2. Adjust the frequency of the signal generator to that of the dominant resonance of the transducer - about 40 KHz. All subsequent measurements will be made at this frequency.

3. Move the detector along the meter stick, and observe that the signal of the oscilloscope passes through several maxima. The displacement of the detector that corresponds to each maximum is **one-half** the wavelength of the standing sound wave. Determine this wavelength by measuring the displacement corresponding to ten successive maxima.

4. Using the oscilloscope's time base setting, measure the frequency of the oscillator (frequency generator).

Experiment (c)
1. Measure the wavelength of sound according to the method described in the Introduction. Use detector displacements which are at least five wavelengths in magnitude.

2. Record the frequency of the wave.

ANALYSIS:

1. Calculate the theoretical prediction of the speed of sound using eq.(5).

2. Use eq.(7) and the data measured in Experiment (a) to calculate the speed of sound in air. The speed of sound in brass is 3480 m/s.

3. Calculate the speed of sound according to the wavelength and frequency measured in Experiment (b).

4. Calculate the speed of sound according to the wavelength and frequency measured in Experiment (c).

5. Compare your experimental determinations of the speed of sound with the expected value as obtained from eq.(5).

EXPERIMENT 3

DISPERSION OF GLASS

OBJECTIVE:

To determine how the refractive index of a prism depends upon the wavelength of the incident light.

REFERENCES:

<u>Physics</u> by Tipler and <u>Experimental College Physics</u> by White and Manning.

EQUIPMENT:

Spencer spectrometer, helium spectral source, low power lamp and a glass prism.

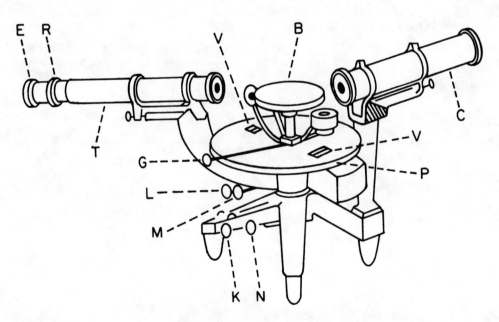

Fig.1. The Spencer spectrometer.

T: Telescope
C: Collimator (rigidly attached to the base)
B: Table (rotating)
P: Platform
V: Vernier (associated with the platform)
G: Clamping Screw (locks Table B)

L: Telescope Locking Screw
M: Fine Angular Adjustment Screw for Telescope
K: Fine Angular Adjustment Screw
N: Fine Angular Adjustment Screw
S: Width Adjustment for Slit Aperture on Collimator

INTRODUCTION:

The Spencer spectrometer is illustrated in fig. 1. The telescope and the platform can be independently rotated around a common vertical axis. Their relative orientation can be indicated by reading the fiducial marks on the divided circular scale which rotates with the telescope. Adjustments have already been made to the spectrometer before it was placed in the laboratory. Do not make major changes. Follow the adjustments in the order listed.

> **Use great care not to force any movement or to clamp any part too tightly.**

To adjust the spectrometer, several steps must be taken:

1. The Eyepiece

Look through the telescope, with your eye relaxed, at a blank sheet of paper and slide the eyepiece gently in and out until the crosshairs are sharply defined. If you wear spectacles, this should be done with them on. Observe that the crosshairs are fixed in the barrel of the telescope and cannot readily be reoriented.

2. The Telescope

Sight upon a distant object (so that parallel rays enter the telescope) and turn the focus ring R to bring the object into focus with the crosshairs. Move your eye laterally to and fro and make fine adjustments to the focus until there is no relative motion between the 'object' and the crosshairs. When this happens, the object image and the crosshairs are in the same plane and have no parallax between them.

3. The Collimator

Align the telescope and the collimator and view the slit to be illuminated with the low power lamp. Turn ring S until the image of the slit is a fine line. Focus this upon the crosshairs by sliding the slit housing in or out, using the no parallax method mentioned above. Rotate the housing without upsetting the focus until the slit is vertical.

> **All adjustments once made should not be disturbed. If they are accidentlally altered the entire procedure should be repeated step by step.**

The light rays, as they pass through the correctly adjusted spectrometer are shown in fig 2.

To read the spectrometer:

1. The instrument scale must be read at the desired vernier setting. In the present case, only one of them needs to be read.

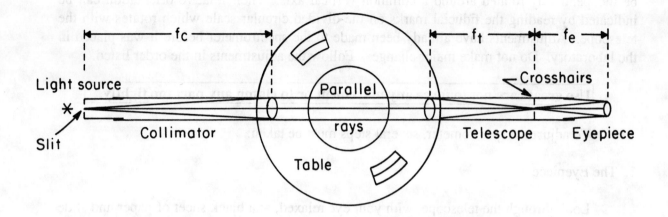

Fig. 2. Light rays passing through correctly adjusted spectrometer.

2. The least count Q of the vernier needs to be known to read the scale properly. Here $Q=\theta/n$, where θ is the smallest main scale division and n is the number of small scale divisions. In the present case, $\theta=1/2$ degree and $n=30$. Therefore, the least count is one minute of arc. This represents the lower limit of error in reading the scale.

3. When taking a reading, always record the main scale setting just before the fiducial mark.

4. Add to this the number of vernier divisions counted to the point where a vernier mark coincides with a main scale mark.

THEORY:

The refractive index of a material is not a constant but depends on the wavelength of the incident radiation. This phenomenon is known as dispersion and it explains such occurrences such as rainbows and the separation of white light into its spectral colors by means of a prism. To measure the index, n, as a measure of wavelength, λ, we shall make use of an expression involving the minimum deviation, ψ_0, of a prism of angle α,

$$n = \sin \tfrac{1}{2}(\alpha+\psi_0) / \sin \tfrac{1}{2} \alpha \ . \tag{1}$$

By illuminating such a prism with a spectral line source which emits visible radiation of characteristic and known wavelengths and then measuring the minimum deviation corresponding to each wavelength, we may establish the dependence of n upon λ.

PROCEDURE:

1. Determine the Prism Angle.

Gently clamp the prism to the table with the frosted face adjacent to the clamping post, as shown in fig. 3. Lock the platform and turn the table until the refracting edge of the prism points towards the collimator. Lock the table. Illuminate the slit with the low power lamp. An image of the slit can readily be seen with the naked eye on reflection from either face of the prism. Look through the telescope and rotate it until one of the images is seen

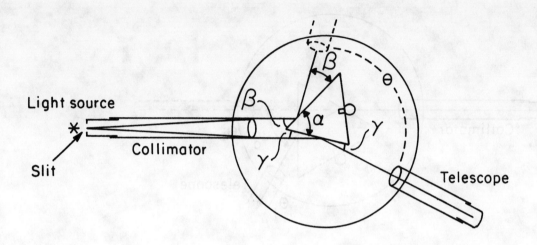

Fig.3. Set-up for measuring the prism angle.

near the crosshairs. Lock the telescope and by means of the fine adjustment screw bring one edge of the image into coincidence with the intersection of the crosshairs. Record the reading using the vernier. Unlock the telescope, swing it over to the other side and repeat the above procedures for the other image. The difference between the two readings is twice the angle of the prism.

> NOTE: Use the same vernier each time. If the divided circle passed through zero as the telescope was swung over, take this into account.

2. Determine the Angle of Minimum Deviation.

Unlock the table and orient the prism so that light from the helium lamp passes through the prism and suffers dispersion as shown in fig. 4. The spectral lines should be clearly visible with the unaided eye.

Place the telescope in position and select a particular line for observation. (If necessary, adjust the location of the lamp so that the spectrum is as bright as possible.) Make sure the platform is locked. Rotate the prism table slowly and follow the selected line, keeping it centered on the crosshairs of the telescope. At some point, the line will be seen to reverse its motion. (If it does not, rotate the prism table in the opposite sense). <u>This is the position of minimum deviation for that particular wavelength</u>. Determine this position carefully by using the fine adjustment screw on the telescope. If this has been accurately done, small oscillations of the prism table will cause the spectral line to move away from the crosshairs in one direction only. Record the reading of the vernier. Redetermine this position, for the same line, several times and calculate the average value, θ(j). This procedure should be repeated for all of the spectral lines that can be seen clearly. Remove the prism and rotate

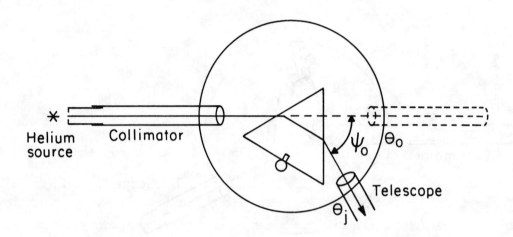

Fig. 4 Set-up for finding angle of minimum deviation.

the telescope until it is directly opposite the collimator. Line up the crosshairs with the slit and record the reading. Do this several times and calculate the average value θ(0). The angle of minimum deviation $\psi_0(j)$ for the jth spectral line is $|\theta(j) - \theta(0)|$.

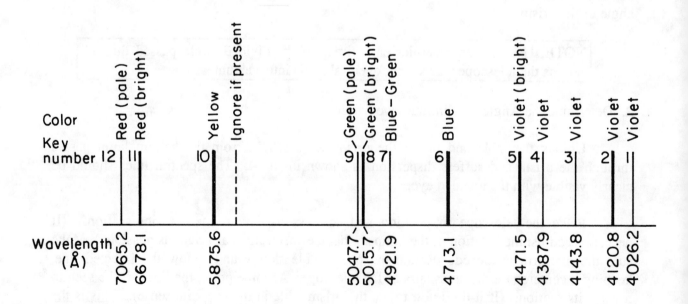

Fig. 5. Color spectrum for He lamp, with key numbers.

DATA COLLECTION:

The data should be collected in a systematic fashion so that the subsequent computer calculation may be easily performed. Each line in the spectrum is assigned a key number as shown in fig. 5 which also includes wavelength and color. It is important that each line be identified by the correct key number. The data should be formulated under the following headings:

Key Number   Color $\theta(j)$ $\theta(0)$ $\Psi_0(j)$

ANALYSIS:

Eq. (1) could now be used to calculate n for all the wavelengths and the results plotted as a graph of n vs. λ. However, it is often more convenient to obtain a functional relationship between n and λ such as

$$n(\lambda) = B_1 + B_2/\lambda + B_3/\lambda^2 + ... \qquad (2)$$

where B_1, B_2 and B_3 are constants. The series has been truncated at the third term since higher order terms are not warranted by the accuracy of this data. Three values of n from eq.(1) and their corresponding wavelengths could be substituted into eq.(2) to obtain three equations from which the unknown B's could be calculated. However, it is more desirable to use all the experimental data as in the method of least squares. Since a manual application of this method of least squares is somewhat tedious, a computer program has been written to do this lengthy calculation and appears on the following pages.

THE COMPUTER PROGRAM:

You will need to run two programs on a PC. The first program, called PRISM, will read your data from a file which you must write into the memory of your computer. It will then calculate the three constants, B_1, B_2 and B_3 from eq. (2), using the least squares method. It will print a table for your report and it will also write a file called PRISM.OUT. The file PRISM.OUT will be read by the plotting program.

The plotting program, called PLOT, will draw a graph of index vs. wavelength for your lab report.

The following is the procedure for using the PC's in the Penn State Academic Computing Labs. From time to time the Academic computing staff may revise their programs so the instructions in this manual may not be quite accurate. Your instructor will run the program before the week of your lab and tell you about any changes.

1. Take your data to a computer lab and ask for a PC. Give the operator your ID card, and he/she will give you a diskette and tell you which machine to use.

2. Insert the diskette into the left hand slot of the computer.

3. If the computer is on, turn it off and wait a few seconds before turning it on again.

4. You will then be asked to type in your name to identify your output at the printer; do this now and then press **ENTER**.

5. Next you will get the menu screen. Exit this by pressing the appropriate selection. (In this case press **X**.)

6. You are now in the DOS (Disk Operating System). A prompt should appear that looks like D:\> or C:\>

7. You now type in:
   copy con prism.dat
   and then hit the **ENTER** button.

8. The cursor will move down to the next line. Now type in your data <u>exactly as you see the sample data below</u>.

   *Leading blanks or zeros are necessary for numbers less than ten.
   *The third line is the number of data (12) and the angle of the prism (49 43). If you have a different number of data points the 12 would be changed accordingly.
   *The inputed data does not have to be in perfect order as the program will sort it out.

9. Sample Input:

   ```
   Michael D'Alesandro
   Prism no. 19
   12 49 43
    1 45 07
    2 44 47
    3 44 34
    4 43 42
    5 43 47
    6 42 44
    7 42 27
    8 42 17
    9 42 14
   10 41 10
   11 40 31
   12 40 16
   ```

   *<u>Remember to include a space in front of numbers less than 10.</u>

10. After you have typed in your last file followed by hitting the **ENTER** key, hold down the **Ctrl** key and at the same time type **Z**. Then hit the **ENTER** key.

11. The C:\> or D:\> prompt will then appear again.

12. If you wish to see your file, type **type prism.dat**

13. You are now ready to run the program called PRISM. Next to the C:\> or D:\> prompt, type **CLASS PHYS204 PRISM,** followed by hitting the **ENTER** key.

14. After a few seconds, the message: Stop Program Terminated will appear on your screen. Your output is then spooled to the printer where you can go and pick it up. If you get any other message, then your data file has been incorrectly typed.

15. Next run the PLOT program. Type **CLASS PHYS204 PLOT** followed by hitting the **ENTER** key.

16. First you will see some instructions. Read them and then go to the next screen.

17. You will want what is called Screen 2 for the graph so choose the default option.

18. The program will then prompt you for maximum and minimum values for the index. Look at the output of the program you just ran to get this information.

19. A graph will come on the screen when you press **ENTER** after inputting n minimum.

20. If you like this graph, send it to the printer by depressing the **SHIFT** key and, at the same time, pressing the **Print Screen** key. Nothing will happen for over a minute so be patient!

21. If you do not like the graph, push any key and enter new maximum and minimum values.

22. If the paper perforations go through the graph, ask the operator to fix the printer and then repeat the graphing procedure.

23. Press **ESC** button to exit this program.

You should use a pen to letter the ordinate and abscissa scales on the graph. Calculate the values of the optical constants n_D and v which are given by the formulas below.

$n_D = n(589.3 \text{ nm})$
$v = (n_D - 1)/[n(486.1) - n(656.3)]$

These constants are used by lens designers. 589.3 nm is the wavelength of the yellow sodium line that is emitted by sodium vapor lamps or by salt or glass heated in a Bunsen flame. 486.1 and 656.3 nm are wavelengths of blue and red light emitted by atomic hydrogen. The quantity v, called the Abbe constant, is used in the design of achromatic lenses. Your instructor knows the values of these constants for the type of glass from which your prism is made.

Verify eq. (1), with the assumption that the condition of minimum deviation is that the light ray travels symmetrically through the prism.

EXPERIMENT 4

THE MICHELSON INTERFEROMETER

OBJECTIVE:

To measure the refractive index of air using the Michelson interferometer.

REFERENCES:

<u>Fundamental of Optics</u> by Jenkins and White, <u>Optics</u> by Hecht and Zajac or <u>Geometrical and Physical Optics</u> by Longhurst.

EQUIPMENT:

He-Ne laser, component table and adjustable mount with front surface plane mirror all on 1/2 meter optical bench, beam splitter and 8 mm focal length diverging lens each with magnetic mount, gas cell with vacuum pump and gauge, meter scale and screen.

INTRODUCTION:

Interference effects involve the superposition of two or more coherent beams and they may be conveniently grouped into two fundamentally different classes. The first involves **division of the wavefront** into two or more parts which are then subsequently recombined after having travelled different paths as in the multislit interference experiment. The second is based on **division of amplitude** whereby the primary wavefront is divided by partial reflection into two or more wavefronts each maintaining the original width but having reduced amplitudes. The Michelson interferometer is an ingenious and important example of this second class.

The Michelson interferometer has the basic configuration shown in fig. 1. Light from an <u>extended source</u> (this may be a ground glass screen illuminated by a discharge lamp) falls upon a beam splitter whereupon it is amplitude divided at the half silvered surface into two beams of approximately equal intensity, A and B, which travel respectively towards the front surface plane mirrors M_1 and M_2. After reflection the beams return to the beam splitter, with A being transmitted and B being reflected toward the screen where an interference pattern may be observed. The unsilvered compensating plate, if present, should be identical to the beam splitter and parallel to it so that the optical path <u>in glass</u> is the same for the two beams. This is not essential for producing fringes from a monochromatic source but it is indispensable when white light is used. (WHY?) If an extended monochromatic source is employed the fringes may either be circular if the mirrors M_1 and M_2 are exactly perpendicular to each other or <u>almost straight</u> and localized if M_1 and M_2 are not exactly orthogonal. In the first case, if the difference ℓ between the distances from the mirrors M_1 and M_2 to the beam splitter is more than a few centimeters the fringes will be very closely spaced and difficult to observe with the unaided eye.

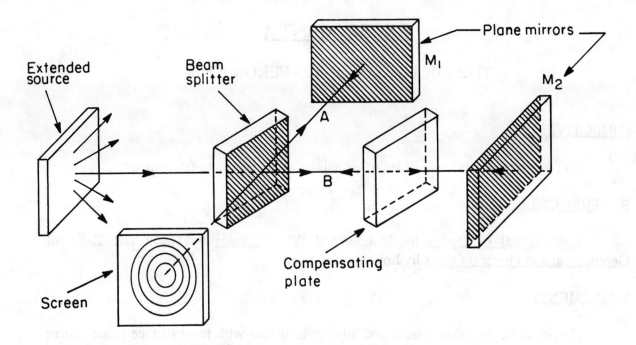

Fig.1. The Michelson interferometer.

Circular Fringes. As indicated above these are produced by monochromatic light when the mirrors M_1 and M_2 are exactly perpendicular to one another. The formation of these fringes can be understood from fig. 2. Here, M_2' is the virtual image of M_2 as seen by reflection in

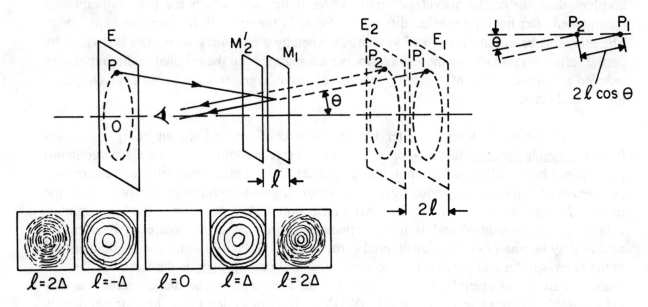

Fig.2. Circular ring formation.

the beam splitter and clearly M_2' and M_1 are parallel. Also, owing to the several reflections that occur in the actual interferometer, the extended source is effectively at position E as shown. To the eye, a single point P on this source will form, on the planes E_2 and E_1 respectively, two virtual images P_2 and P_1 which act as coherent point sources separated by a distance 2ℓ. The path difference between the two rays entering the eye is $2\ell \cos \theta$. The condition [1] for observing constructive interference is, where n is the index of refraction of air,

$$2\ell n \cos \theta = m\lambda, \quad m = 0, 1, 2, \text{-----INTERFERENCE MAXIMA} \quad (1)$$

Clearly, if this condition is satisfied by a source point P it must equally well be satisfied by all other points which lie on a circle of radius OP in E where O is on the axis of the system. For given m, λ and ℓ, the angle θ is constant so the maxima will be in the form of concentric circles around the axis. A fringe of particular order m will be observed at angle θ given by $\cos \theta = m\lambda/2\ell n$; if M_2' is now moved towards M_1, ℓ decreases and θ must become smaller. Therefore the fringes shrink towards the centre with the **highest** order one disappearing when ℓ decreases by $\lambda/2n$. The remaining fringes get wider as more fringes disappear at the center until, when $\ell = 0$, the central fringe has spread out to fill the entire field of view, (the path difference is now zero for all angles, θ, of incidence). If M_2' is moved still more so that in effect it passes through M_1, new broadly spaced fringes appear from the center. These become more closely spaced as ℓ continues to increase. Note that in fig. 2 the rays entering the eye are parallel so that the eye must be focussed at infinity.

Localized Fringes. Suppose that one of the mirrors, M_2 say, is tilted slightly so that, as shown exaggeratedly in fig. 3, M_2' and M_1 are inclined at a small angle thereby enclosing a wedge shaped space between them. A point P on the extended source will form two virtual images P_2 and P_1 on the image planes E_2 and E_1 respectively. Thus P_2 and P_1 act as coherent (virtual) sources giving rise to an interference pattern at near the mirror M_1. The path difference, $ab + bP' - aP'$, between the two rays arriving at P' may be calculated in the following manner.[2] If points c and d are the images of P' in M_2' and M_1 respectively then c and d both lie on the direction of the original ray Pab. Geometrical considerations indicate that $OP' = Oc + Od$, therefore P', c and d lie on a circle of radius $r = OP'$. The path difference may then be rewritten as $ab + bd - ac = 2r \sin \phi$ where ϕ is the angle of wedge. There will be various source points P for which the associated reflected rays will pass through P'; for all such points the path difference remains constant. Consequently the constructive interference condition may be expressed as

$$2rn \sin \phi = m\lambda, \quad m = 1, 2, 3 \text{-----INTERFERENCE MAXIMA} \quad (2)$$

Further, if the angle between M_2' and M_1 is not too small the distances of P' from M_1 are not great, so that the system of fringes is formed in the vicinity of the mirror M_1. These localized fringes are **almost** straight – especially if the distance ℓ between the "mirrors" is

---

[1] One of the beams actually undergoes internal reflection at the beam splitter and there may be a phase change of π depending upon whether the beam splitter is silvered or not. Will this have any bearing on eq. (1)?

[2] <u>Light</u> by R. W. Ditchburn, 2nd edition p. 136 Blackie & Sons Ltd. London (1963).

localized fringes are **almost** straight – especially if the distance ℓ between the "mirrors" is quite small – and are oriented parallel to the line of intersection of M_2' and M_1. If ℓ is appre-

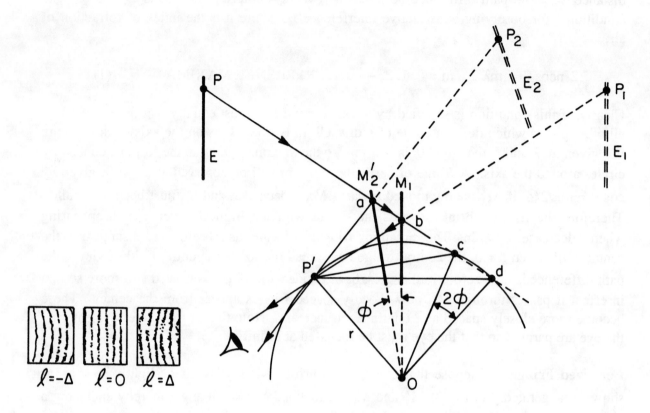

Fig. 3 Tilted mirrors.

ciable the fringes will exhibit some curvature (always convex, towards the apex of the wedge) which arises from matters of variation with angle in planes perpendicular to that of fig. 3. Nevertheless, as ℓ is varied the fringes move across the field of view with a fringe passing some reference mark each time ℓ is changed by $\pm\lambda/2n$. Note that the spacing of the fringes is independent of ℓ, being given by $\lambda/2n \sin \phi$, as is evident from eq. (2).

The Effects of the Index of Refraction of Air. The index of air may be measured by installing a gas cell in arm A as illustrated in fig. 4. (Disregard the presence of the diverging lens for the moment - its function will be described later). Suppose that the gas cell, the interior length of which is t, is filled with gas having a density ρ_1 and index of refraction n_1 corresponding to some pressure P_1. Then an interference pattern will be visible assuming that the interferometer is in adjustment for either circular or straight fringes. Now let the pressure in the cell be changed to P_2, corresponding to a gas density ρ_2 and an index of refraction n_2. The optical path length in the arm A will have changed by $2(n_2-n_1)t$ and consequently a number Δm (which may be fractional) of fringes will have moved past some conveniently chosen reference mark in the middle of the field of view. Clearly we may write

$$2(n_1-n_2)t = \Delta m \lambda \tag{3}$$

thereby relating the change in refractive index of the gas to the number of fringes that have moved past the mark.

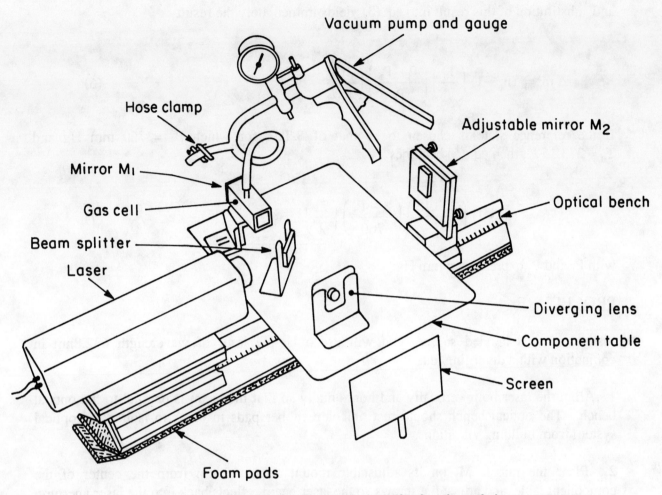

Fig.4. Set-up for measuring the index of refraction of air.

For gases at reasonable temperatures and pressures it may be assumed that the ideal gas law, $PV = nRT$, and the Dale-Gladstone relation, $(n-1)/\rho = C$ where C is a constant, are valid. From the latter we may deduce that

$$n_1 - n_2 = (\rho_1 - \rho_2)C, \text{ with } C = (n_0 - 1)/\rho_0 \qquad (4)$$

where n_0 and ρ_0 are, respectively, the index of refraction and the density of the gas in some arbitrarily chosen reference state having pressure P_0 and absolute temperature T_0. Use of the ideal gas law leads to the relationship $\rho = (M/R)(P/T)$ where M is the molecular weight and R is the gas constant, so that at constant temperature $\rho_1 - \rho_2 = (M/R)(P_1 - P_2)/T$. Eq. (4) then takes the form

$$n_1 - n_2 = (n_0 - 1)\left(\frac{T_0}{T}\right)\left(\frac{P_1 - P_2}{P_0}\right), \tag{5}$$

and substitution of this result into eq. (3) yields immediately the result

$$\Delta m = (n_0 - 1)\left(\frac{T_0}{T}\right)\left(\frac{P_1 - P_2}{P_0}\right)\left(\frac{2t}{\lambda}\right) \tag{6}$$

If the reference state is chosen to be that of S.T.P. for which $P_0 = 760$ mm Hg and $T_0 = 273.15$ K, then eq. (6) becomes

$$\Delta m = (n_0 - 1)\left(\frac{273.15}{T}\right)\left(\frac{P_1 - P_2}{760}\right)\left(\frac{2t}{\lambda}\right) \tag{7}$$

with P_1 and P_2 expressed in mm Hg.

PROCEDURE:

Instead of an extended source, you will use a He-Ne laser of wavelength 632.8nm in conjuction with a diverging lens.

1. Align the laser both vertically and horizontally so that the beam is parallel to the optical bench. The optical bench should rest on foam rubber pads in order to isolate the optical system from building vibrations.

2. Place the mirror M_2 on its adjustable mount about 15 cm from the center of the component table, and adjust the screws so the laser beam reflects back into the laser aperture. Thus M_2 is perpeindicular to the laser beam.

3. At the proper place on the component table, accurately set the beam splitter at 45° so that the reflected beam is at 90° with respect to the incident beam.

4. Connect the gas cell to the table using the screw and wing nut provided. Adjust it as necessary to insure that *its* incident and return rays are coincident as determined by the coalescing of two spots visible on the rear surface of the beam splitter. This alignment must be made very carefully, otherwise the interference fringes will be too closely spaced to see. M_1 and M_2 should nearly be perpendicular.

5. You should see a pair of bright spots, as well as other fainter pairs of spots, on the screen. Readjust M_2 until the brightest pair of spots coincide. M_1 and M_2 are now accurately adjusted to a right angle.

6. Insert the lens of focal length – 8mm as illustrated in fig.4. Interference fringes should be evident on the screen. If there are only two large spots, reorient M_2 until the spots are confluent and fringes appear.

7. The fringes should be circular. If so, slide M_2 in and out along the optical bench and observe how the fringe apearance and fringe spacing change as described in the Introduction. Leave M_2 at a position which yields a few broad circular fringes in the field of view. More than likely, however, M_1 and M_2 will form a thin wedge and the fringes will be almost straight. You may wish to reorient M_2 and/or the gas cell until you obtain suitable, widely spaced fringes. Note how the curvature of these fringes alters as M_2 is moved in or out along the bench.

8. Pump out the gas cell a little to be sure they change; if they do not, you picked the wrong spots in step (5).

9. Put a reference mark on the screen to help you record the fringe shift.

10. Pump out the cell, and adjust the leak rate with the hose clamp such that after a full pump down, atmospheric pressure will be reached in about 2 minutes.

11. Pump out the cell again.

12. Beginning with any observed pressure P_1, count the number of fringes Δm that pass the reference mark by the time the pressure reaches some value P_2.

13. Repeat step (11) 4 more times, but choose different values of P_1 and P_2 for each trial. As you take the data Plot Δm vs. $(P_1-P_2)/760$ on linear graph paper.

14. Carefully measure t, the inside length of the gas cell.

15. Measure the room temperature, T.

ANALYSIS:

1. From the slope of the line for Δm vs. $(P_1-P_2)/760$, determine the index of refraction of air, n_0.

2. Compare the experimental value of n_0 to the expected value.

3. How can you account for the number of spots seen on the screen after step (3)? What about step (4)?

4. The gauge reads gauge pressure, not absolute pressure. Does this affect your calculations?

5. Which measurement contributes the most error in the experiment? What suggestions do you have for reducing error?

EXPERIMENT 5

INTERFERENCE - NEWTON'S RINGS

OBJECTIVE:

To study the formation of the interference pattern known as Newton's Rings and to deduce the wavelength of the monochromatic light producing them.

REFERENCES:

<u>Geometrical and Physical Optics</u> by Longhurst, <u>Fundamentals of Optics</u> by Jenkins and White or <u>Optics</u> by Hecht and Zajac.

EQUIPMENT:

Plano convex lens of long focal length, glass plates, microscope with micrometer eyepiece, sodium lamp, optical bench and accessories, plane mirror, spherometer and vernier caliper.

INTRODUCTION:

If the convex side of a plano-convex lens of long focal length is placed on a flat glass plate a thin air film of varying thickness is obtained - see fig. 1. The film is symmetrical about the point of contact. If the system is suitably illuminated from above and observed in reflection a set of interference fringes is seen. The fringes are in the form of concentric circles and are known as Newton's Rings. They result from the interference of light that is reflected from the top and the bottom of the air film. A relationship can be derived between

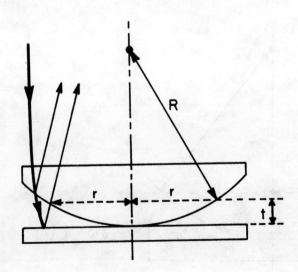

Fig.1. Newton's ring geometry.

the radius r of the rings, the wavelength λ of the light and the radius of curvature R of the convex surface. If t is the thickness of the air film at a distance r from the point of contact, then

$$t = \frac{r^2}{2R - t} \quad . \tag{1}$$

Since usually R>>t, eq.(1) may be rewritten as

$$t = \frac{r^2}{2R} \quad . \tag{2}$$

The index of refraction of air is 1. Recalling that there is a phase change of π on reflection at the lower surface, the condition for destructive interference is

$$2t = m\lambda, \quad m = 0, 1, 2 \ldots \tag{3}$$

On combining eqs. (2) and (3) it is readily seen that $r^2 = (\lambda R)m$, so that the diameter D of the dark rings is given by

$$D^2 = (4\lambda R)m, \quad m = 0, 1, 2, \ldots \tag{4}$$

Eq. (4) predicts that the centre of the pattern (m=0) is a dark spot in agreement with the experimental observation. This provides confirmation of the existence of the phase change of π on reflection at a denser medium. A plot of D^2 against the ring number m will yield a straight line of slope $4\lambda R$ so that if the curvature of the convex surface is known, the wavelength may be determined. The effect of distortion over the region of contact is merely to alter the intercept on the D^2 axis. This is easily seen since the thickness of the air film becomes

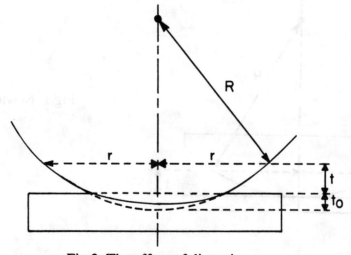

Fig.2. The effect of distortion.

$$t = \frac{r^2}{2R} - t_0 \qquad (5)$$

where t_0 is the distortion as defined in fig. 2. Consequently eq. (4) is now to be rewritten as

$$D^2 = (4\lambda R)\, m + 8t_0 R \qquad (6)$$

PROCEDURE:

1. Set up the apparatus shown in fig.3. Set the glass plate with the black side down, and place the lens on the plate as shown. Above this, arrange another glass plate at a 45° angle so that the light from the sodium lamp is normally incident upon the lens and lower plate. The interference rings which occupy a small region around the center of the lens may then be viewed with the microscope.

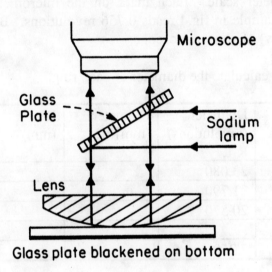

Fig.3. The apparatus used to view interference rings.

2. Adjust the eyepiece of the microscope until the hairline is seen clearly.

3. Look for the interference pattern, and bring it into focus on the hairline by sliding the microscope up or down. You may need to reposition the lens so that the rings are in the middle of the field of view. If the central fringe is not dark, there is some dirt on the contacting surfaces which must then be cleaned.

4. Move the hairline to the left of the 10th ring. Then reverse the direction of the screw, and set the hairline exactly on the 10th ring. This decreases error due to backlash of the screw. The hairline should bisect the thickness of the fringe as shown in fig.4.

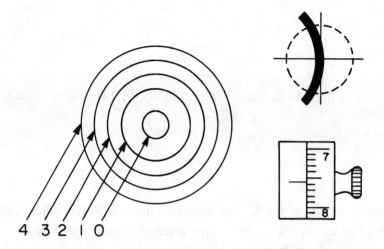

Fig.4. Method for measuring the ring diameters.

5. Record the reading of the micrometer scale. Each mark on the micrometer screw represents 0.01 revolution. Thus the example in fig.4 reads 0.756 revolutions. Be sure to record the calibration constant (in mm/rev) of your microscope.

6. Make a data table similar to table 1 to calculate the diameter of each ring.

Ring number, m	Left Micrometer Reading, L	Right Micrometer Reading, R	Diameter (revolutions) R-L	Diameter (mm)	D^2 (mm)^2
10	0.270	23.350	23.080		
9	0.930	22.720	21.790		
8	1.555	22.125	20.570		
...	...	...	...		
1	8.175	15.425	7.250		

Table 1. Sample Data Table.

7. Moving from left to right, record the reading of the micrometer scale for the 9th ring.

8. Similarly, record the readings for each ring until you reach the 1st one. Be sure to count each complete revolution of the micrometer screw, because each revolution must be added to the scale reading in a cumulative manner as you proceed systematically across from one side to the other.

9. Cross the central dark spot (m=0), but do not measure its diameter. Go directly to the 1st ring on the right, and read the micrometer scale.

10. Continue making readings of each ring until you reach the 10th ring. **Always turn the micrometer screw in the same sense as you set the hairline, for this will avoid backlash.**

11. You will need to measure the radius of the lens. You can do this by two methods — the optical method and the mechanical method.

Optical Method. An optical method for finding R is illustrated in fig. 5. Consider a ray AB from an object at A so placed that the refracted ray travels along BC, the normal to the curved surface of the lens. At this surface part of the incident light is transmitted and part is reflected. The reflected portion returns along the path CBA and forms a real image at A. The transmitted portion, however, appears to come from E which is the center of curvature of the surface, so that **if one looks through the lens the image of A appears to be at E.** The (negative) image distance I which is equal to the negative of the radius of curvature of the surface may then be calculated using

$$\frac{1}{O} + \frac{1}{I} = \frac{1}{f} \tag{7}$$

if the distance O and the focal length f are known.

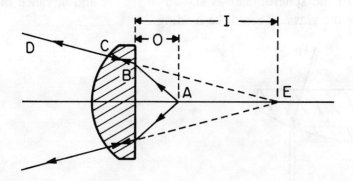

Fig.5. Determination of O.

1. Set up the lens and the light source on the optical bench. The light source has a central object in the form of a cross and is surrounded by a screen in the same plane.

2. Using the curved surface of the lens as a mirror, adjust the screen to lens separation until a sharp image of the cross appears on the screen.

3. Measure the screen to lens distance; this is the object distance O.

4. The focal length f can be found by measuring the image distance with an infinite object distance, or by the method of autocollimation. You will use autocollimation to determine f.

5. Set up the lens such that the plane side is facing the light source, as in fig. 6.

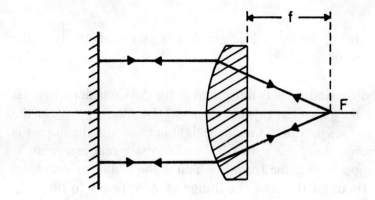

Fig.6. Determination of f.

6. Place a plane mirror behind the curved side of the lens, and adjust the light source until its own image falls on itself. The lens to screen distance is the focal length f. Record this measurement.

Mechanical Method

1. Examine the spherometer, and determine the least count of the scale.

2. Place the lens on the three legs of the spherometer as shown in fig. 7, and advance the screw until it just makes contact with the glass. Record the reading.

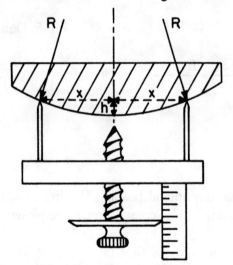

Fig.7. Spherometer.

3. Replace the lens with a flat glass plate. Advance the screw (in the same direction to avoid backlash) until it touches the surface, and record the reading.

4. Determine the sagitta h, the difference between the two readings.

5. Repeat this measurment several times, and calculate the average value of h.

6. Place the spherometer on a sheet of paper, and make an impression of the three legs and the screw point as in fig. 8. Remember, you never get a second chance to make a good first impression!!!!

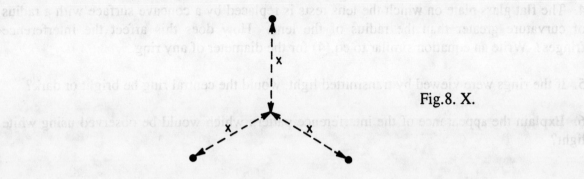

Fig.8. X.

7. With a vernier caliper, measure the distance X from the center point to the point of each leg.

8. Calculate the average value of X.

ANALYSIS:

1. Make the necessary calculations to complete your data table.

2. Using eq.(7), the object distance, and the focal length, calculate the radius of curvature. (R will be the negative image distance in this case.)

3. Using eq.(9), calculate the radius of curvature.

4. Plot D^2 vs. m.

5. From eq.(6) calculate the wavelength of sodium light λ and the distortion t_0.

6. Compare your answer for λ with what you would expect. (Note λ = 589 nm for the sodium D lines.)

7. Estimate the uncertainty of λ, and show how you arrived at this value.

Questions

1. In fig.6, why is the object distance not equal to R?

2. In fig. 3, a drop of distilled water is placed on the glass plate and then the lens is placed on it. Are the interference fringes affected? If so, is the diameter of the 10th ring larger or smaller than before? What is its diameter if the refractive index of water is 4/3?

3. Would the rings be larger or smaller for red light than for the sodium light?

4. The flat glass plate on which the lens rests is replaced by a concave surface with a radius of curvature greater than the radius of the lens. How does this affect the interference fringes? Write an equation similar to eq.(4) for the diameter of any ring.

5. If the rings were viewed by transmitted light, would the central ring be bright or dark?

6. Explain the appearance of the interference pattern which would be observed using white light?

EXPERIMENT 6

DIFFRACTION OF LIGHT

OBJECTIVE:

To investigate the Fraunhofer diffraction of monochromatic light by various combinations of linear apertures.

REFERENCES:

Fundamentals of Optics by Jenkins and White, Optics by Hecht and Zajac or Geometrical and Physical Optics by Longhurst.

EQUIPMENT:

He-Ne laser, optical bench, mounts, Ealing slit mosaic, micrometer with travelling eyepiece, lamp, meter stick and adhesive tape.

INTRODUCTION:

When a beam of light passes through a narrow aperture it spreads out into the region of the geometrical shadow. Such an effect is known as diffraction and it may be qualitatively classified into two types. (i). When both the incident and diffracted waves are plane we have FRAUNHOFER diffraction. (ii). When the curvature of either or both of the two waves is appreciable we have FRESNEL diffraction. We shall only discuss Fraunhofer diffraction as it is mathematically much simpler to analyse. It may be conveniently realized in the laboratory by using a collimated parallel beam, as from a laser, as the incident wave and by placing the viewing screen at a large distance from the diffracting aperature. (Alternatively, a positive lens may be used to collect parallel rays and to focus the diffracted beam onto a screen at the focus of the lens.)

Diffraction by a single slit. The intensity, as a function of angle θ, for the diffraction pattern from a long single slit of width a is given by

$$I = I_0 \left[\frac{\sin\left(\frac{\pi a \sin\theta}{\lambda}\right)}{\frac{\pi a \sin\theta}{\lambda}} \right]^2 = I_0 \left(\frac{\sin\alpha}{\alpha}\right)^2 \quad (1)$$

with

$$\alpha \equiv \frac{\pi a \sin\theta}{\lambda}, \quad (2)$$

where λ is the wavelength of the light used and I_0 is the intensity at $\alpha=\theta=0$, corresponding to the principal maximum. See fig. 1. It is important to note that the diffraction pattern is

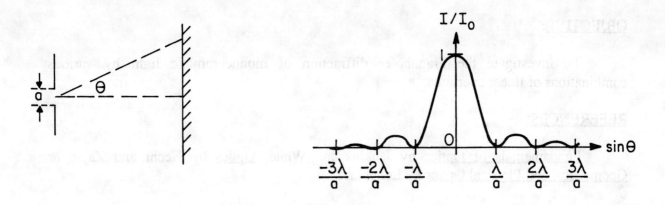

Fig.1. The single slit diffraction pattern.

spread out in a direction which is perpendicular to the length of the slit. The positions of the minima of intensity are found from the condition $\alpha = \pm\pi, \pm 2\pi, \pm 3\pi, \ldots$ which yields the familiar formula

$$\sin\theta = \pm n\frac{\lambda}{a}, \quad n = 1, 2, 3, \ldots \text{ DIFFRACTION MINIMA} \qquad (3)$$

The locations of the maxima are to be found from the requirement $dI/d\alpha=0$ i.e. $\tan\alpha = \alpha$, a transcendental equation having solutions $\alpha = 0, \pm 1.4303\pi, \pm 2.4590\pi, \pm 3.4707\pi, \ldots$ so that we have

$$\sin\theta = 0, \pm 1.4303\frac{\lambda}{a}, \pm 2.4590\frac{\lambda}{a}, \ldots$$
$$\text{DIFFRACTION MAXIMA} \qquad (4)$$
$$\approx 0, \pm 1.5\frac{\lambda}{a}, \pm 2.5\frac{\lambda}{a}, \ldots$$

Consider, say, the second subsidiary maxima at $\alpha = \pm 2.5\pi$. Since $\alpha = \dfrac{\pi a \sin\theta}{\lambda}$, increasing the slit width requires a decrease in θ because α is constant, so that the pattern shrinks inward towards the central or principal maximum, as it would also do if the wavelength λ were decreased.

Diffraction by a double slit. Consider a diffracting aperture consisting of two long parallel slits each of width a and separated by a distance d. The corresponding intensity is given by

$$I = 4I_0 \left[\frac{\sin\left(\frac{\pi a \sin\theta}{\lambda}\right)}{\left(\frac{\pi a \sin\theta}{\lambda}\right)} \right]^2 \left[\cos\left(\frac{\pi d \sin\theta}{\lambda}\right) \right]^2 = I = 4I_0 \left(\frac{\sin\alpha}{\alpha}\right)^2 \cos^2\gamma \quad (5)$$

with α given as above, and

$$\gamma \equiv \frac{\pi a \sin\theta}{\lambda}. \quad (6)$$

As before, I_0 is the intensity from a single slit at $\alpha=\theta=0$. The factor 4 arises from the fact that the amplitude of the wave is twice what it would be if one slit were covered. The $\left(\frac{\sin\alpha}{\alpha}\right)^2$ "diffraction" term in eq. (5) is recognizable as the intensity distribution for a single slit and serves as the envelope for the "interference" term $\cos^2\gamma$ which describes the interferecnce between the diffracted beams from each slit. The appearance of the intensity as a function of $\sin\theta$ for the case $d = 3a$ is shown in fig. 2. If a is small, the diffraction pattern

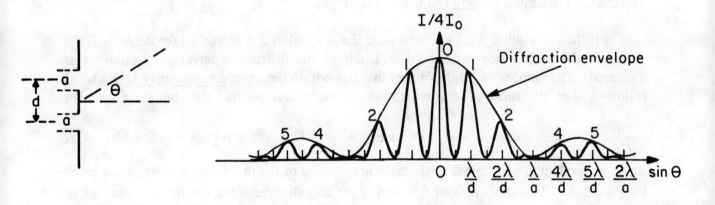

Fig.2. Two slit interference for the case d=3a.

from either slit will be essentially uniform over a broad central region and interference fringes will be evident in that region. The diffraction term will be zero and no light will reach the screen, to be available for interference, at points for which $\alpha = \pm\pi, \pm 2\pi, \pm 3\pi, ...$ In terms of $\sin\theta$ this implies that

$$\sin\theta = \pm n\frac{\lambda}{a}, \quad n = 1,2,3, ... \text{ DIFFRACTION MINIMA} \quad (7)$$

Now, the interference term is also zero at points on the screen for which $\gamma = \pm \frac{\pi}{2}, \pm \frac{3\pi}{2}, \pm \frac{5\pi}{2}, \ldots$ so that the waves from the two slits will be exactly out of phase and cancel regardless of the actual amount of light made available by the diffraction process. This condition may be written as

$$\sin\theta = \pm p \frac{\lambda}{2d}, \quad p = 1,3,5, \ldots \text{ INTERFERENCE MINIMA} \qquad (8)$$

These various minima are clearly illustrated in fig. 2. The exact positions of the maxima are not given by any simple relation. However if variations in the factor $\left(\frac{\sin\alpha}{\alpha}\right)^2$ may be ignored (which is justifiable if the slits are very narrow and if only the central region of the pattern is considered) they will be determined solely by the $\cos^2\gamma$ term which has maxima for $\gamma = 0, \pm\pi, \pm 2\pi, \ldots$ or equivalently,

$$\sin\theta \approx \pm \ell \frac{\lambda}{d}, \quad \ell = 0,1,2,\ldots \text{ INTERFERENCE MAXIMA} \qquad (9)$$

The integer ℓ describes the <u>order</u> of the bright fringes.

If the slit width a is kept constant and the separation d is increased (or decreased) the nature of the interference pattern changes, though the diffraction envelope itself remains unaltered. The number of bright fringes that lies within the central region may be found as follows. Let the ratio of the separation to the width of the slits be some number $c = \frac{d}{a} > 1$. Then, since $\frac{\alpha}{a} = \frac{\gamma}{d}$, we have $c = \frac{\gamma}{\alpha}$. The central region is defined by $-\pi \leq \alpha \leq \pi$ or, equivalently, by $-c\pi \leq \gamma \leq c\pi$ whereas the locations of the interference maxima are given by $\gamma \approx \pm \ell\pi$. Remembering that ℓ is an integer including zero we see that the number of complete bright fringes will be given by $2c-1$ provided that c is rounded down to the nearest integral value. As illustrated in fig. 2 for the case c = 3, there are 5 complete fringes within the central region.

It can happen that an interference maximum and a diffraction minimum (zero) coincide at the same value of θ, so that no light is available at that point and the suppressed fringe is said to be a <u>missing order</u>. This occurs when eq. (7) and (9) are equal i.e. when $\frac{d}{a} = \frac{\ell}{n}$, so the condition for missing orders is that $\frac{d}{a}$ be the ratio of two integers. This ratio determines which orders are missing. For example when d/a = 3 as in fig. 2 the orders $\pm 3, 6, 9, 12, \ldots$ are missing.

Diffraction by multiple slits; the diffraction grating. For the case of N slits each of width a and separation d the intensity is given by

$$I = I_0 \left(\frac{\sin\alpha}{\alpha}\right)^2 \left(\frac{\sin N\gamma}{\sin\gamma}\right)^2, \tag{10}$$

with α and γ given above. For N = 2 this reduces to eq. (5) for the double slit. The $\left(\frac{\sin\alpha}{\alpha}\right)^2$ factor is the single slit diffraction envelope as before, while $\left(\frac{\sin N\gamma}{\sin\gamma}\right)^2$ represents the interference term for N slits. This latter term is shown in fig. (3) for the case N = 6, and d = 4a. It is seen that it possesses principal maxima, proportional to $N^2 I_0$, at $\gamma = 0, \pm\pi, \pm 2\pi, \ldots$ For these we have

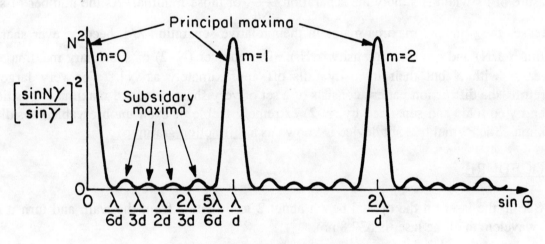

Fig.3 Multiple slit interference.

$$\sin\theta = \pm m \frac{\lambda}{d}, \quad m = 0,1,2,\ldots \text{ PRINCIPAL INTERFERENCE MAXIMA} \tag{11}$$

Note that these maxima correspond in position to those of the double slit (see eq. (9)) but they are more intense by a factor $(N/2)^2$. The relative intensities of the different orders m are in all cases governed by the single slit diffraction envelope (this has not been imposed on the figure). Hence the condition for missing orders is just as described earlier. The zeros of $\left(\frac{\sin N\gamma}{\sin\gamma}\right)^2$ occur at $\gamma = \pm \frac{\pi}{N}, \pm \frac{2\pi}{N}, \ldots \pm \frac{(N-1)\pi}{N}, \pm \frac{(N+1)\pi}{N}, \ldots$ Hence the values

$0, \frac{N\pi}{N}, \frac{2N\pi}{N}, \ldots$ which correspond to the principal maxima are excluded from the sequence. So we have

$$\sin\theta = \pm p\frac{\lambda}{Nd}, \quad p = 1, 2, 3, \ldots (N-1), (N+1), \ldots \text{INTERFERENCE MINIMA} \quad (12)$$

Between consecutive principal mixima – defined by a range of π in γ, or $\frac{\lambda}{d}$ in $\sin\theta$ – there are clearly (N-1) points of zero intensity; the two minima on either side of a principal maximum are twice as widely separated as the others. Between the other minima there are (N-2) low intensity <u>subsidiary maxima</u> which are located approximately midway between the minima. These subsidiary maxima are not all of equal intensity, but fall off as we go out on either side of a principal maximum.

Since a principal maximum is bounded on either side by its adjacent minima, a good measure of its width is simply the separation, $\frac{2\lambda}{Nd}$, of those minima. As the number of slits increases the principal maxima maintain their relative separation λ/d, become ever sharper (width $\propto 1/N$) and brighter (intensity $\propto N^2$). The number (N–2) of subsidiary maxima also increases with N but their intensity falls off approximately as N^{-1}. For very large N therefore, the diffraction pattern consists of a set of very sharp principal maxima at positions given by eq. (11) and separated by (N-2) extremely feeble or even unobservable subsidiary maxima. Such a mulitple slit device is known as a <u>diffraction grating</u>.

PROCEDURE:

1. Set up the laser on the optical bench about 2 to 3 meters from the wall, and turn it on. The wavelength of the laser is 632.8 nm.

Do not look into the laser or point it toward someone's eyes.

2. Tape a piece of linear graph paper to the wall, and orient it in such a way that you will be able to mark the diffraction minima that will appear on the paper.

3. The slit mosaic contains two plates. Plate 1 consists of four sets of double slits with slit spacings as indicated in table 1. Plate 2 consists of four groups containing 1, 2, 3, and 4 slits respectively. Each slit for this plate is approximately 0.05 mm wide, and the separation between multiple slits is approximately 0.11 mm.

Place the mosaic on a mount close to the laser and illuminate the first set of double slits on Plate 1.

double slit number	a (mm)	d (mm)
1	0.035	0.109
2	0.044	0.159
3	0.020	0.261
4	0.046	0.455

Table 1. Plate 1 – Double Slits.

Double Slits

1. Mark the positions of the diffraction minima on the graph paper.

2. Count the number of bright fringes that lie within the central region.

3. Record any missing orders that you might find.

4. Repeat steps (1) through (3) for each of the remaining sets of double slits.

5. Measure the distance L from the mosaic to the screen.

Multiple Slits

1. Remove the mosaic from its mount, and place it under the microscope. Suitable illumination is provided by the lamp. Focus on the group of 3 slits on Plate 2.

2. Accurately measure the slit width a and the slit separation d; if necessary you can refer to Experiment 5 for instructions on how to use the microscope.

3. Using these data, make a sketch similar to fig.3 of the expected diffraction pattern for the central region. Roughly impose the diffraction envelope on the sketch. Label the abscissa (horizontal axis) with the numerical values of $\sin\theta$ at which the principal and subsidiary maxima and the interference minima occur.

4. Place the mosaic on the mount and observe the 3-slit diffraction pattern. Note the presence of the subsidiary maxima.

5. Mark the positions of the various orders on the graph paper.

6. Examine the 4-slit diffraction pattern. Note that it is not of good quality; however, pairs of subsidiary maxima are visible as expected. It is difficult to fabricate a slit system which will behave ideally. In particular, the intensity distribution for a real slit system is often quite different from that predicted by the theory. For this reason, intensity measurements which would yield the slit width a have not been employed in these experiments.

ANALYSIS:

Double Slits

1. For each set of double slits on Plate 1, calculate the positions of the diffraction minima using the approximation that $\sin\theta \approx \tan\theta$ and $\tan\theta = y/L$, where y is the distance from the principal axis to the minima, and eq.(7). Compare these values to the those which were observed.

2. Also calculate the expected number of fringes in the central region for each double slit, and compare these to the observed number of fringes.

3. Do the interference maxima move closer together as the spacing of the double slits increases. What does eq.(9) predict?

Multiple Slits

1. Compare the form of the subsidiary maxima to that of your sketch. Do they agree? If not, state the differences and give a possible explanation.

2. From the positions of the various orders on the screen, calculate the corresponding values of $\sin\theta$, and use eq.(11) to determine an average value of the slit separation d.

3. Compare the calculated value of d with the measured value.

EXPERIMENT 7

DIFFRACTION AND THE WAVELENGTH OF LIGHT

OBJECTIVE:

To examine the diffraction of the light from a He-Ne laser by reflection and to measure its wavelength.

REFERENCES:

<u>Geometrical and Physical Optics</u> by Longhurst, <u>Fundamentals of Physics</u> by Halliday and Resnick or <u>Optics</u> by Hecht and Zajac.

EQUIPMENT:

He-Ne laser, 1/2 meter optical bench with accessories, small plastic scale graduated in 1/50", meter stick, measuring tape.

INTRODUCTION:

Let a parallel beam of monochromatic light, as from a laser, illuminate the end of a uniformly ruled scale at a grazing angle of incidence α as shown in fig. 1. Then in addition to

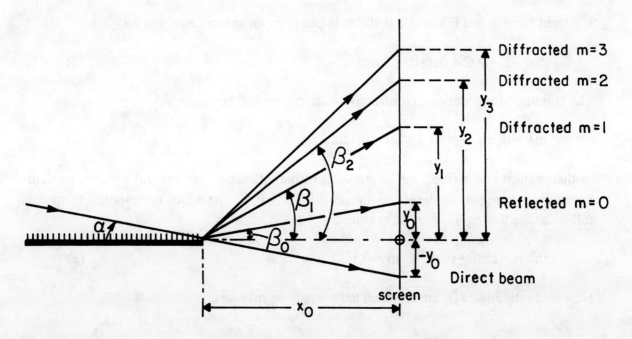

Fig.1. Reflection grating.

the direct and specularly reflected beams, numerous other diffracted beams will be observed on the screen. The precise nature of the pattern will depend upon whether the screen is nearby (Fresnel diffraction) or whether it is far away (Fraunhofer diffraction). Only the latter case will be treated here as it is mathematically simpler.

Consider two parallel rays bB and aA incident at angle α upon two adjacent scale rulings A and E which are distance d apart as illustrated in fig. 2. Rays Cc and Ee, which are

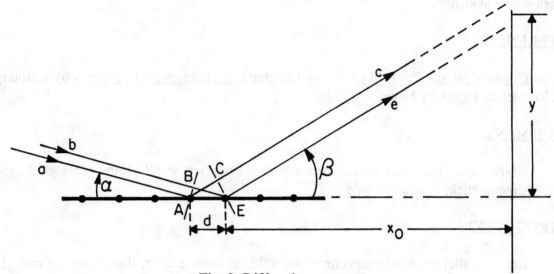

Fig. 2. Diffraction geometry.

scattered from A and E at angle β, differ in phase by an amount Δ, given by

$$\Delta \equiv BE - AC = d\cos\alpha - d\cos\beta .\tag{1}$$

The condition for observing constructive interference is then

$$m\lambda = d(\cos\alpha - \cos\beta_m) \quad m = 0,1,2,...\tag{2}$$

so that, with α and λ fixed, the bright spots of the diffraction pattern will occur at particular angles β_m determined by the choice of m. With $m = 0$, $\beta_0 = \alpha$ which corresponds to specular reflection, so that eq. (2) may be rewritten as

$$m\lambda = d(\cos\beta_0 - \cos\beta_m) \quad m = 0,1,2,...\tag{3}$$

Now since the angles β_m are assumed to be small we may write

$$\cos\beta_m \equiv \frac{x_0}{\left(x_0^2 + y_m^2\right)^{\frac{1}{2}}} = \frac{1}{\left(1 + (y_m/x_0)^2\right)^{\frac{1}{2}}} \approx 1 - \frac{1}{2}(y_m/x_0)^2 + ...\tag{4}$$

where y_m is the vertical height at which the mth bright spot is observed on the screen. Consequently, eq. (3) may be expressed as

$$m\lambda \approx \frac{d}{2x_0^2}(y_m^2 - y_0^2) \qquad (5)$$

On re-arrangement eq. (5) becomes

$$y_m^2 = y_0^2 + \left(\frac{2\lambda x_0^2}{d}\right) m \qquad (6)$$

PROCEDURE:

1. Mount the laser and the graduated scale upon their respective stands on the optical bench so that the incident light just grazes the end of the scale as in fig.1. Be sure that some of the beam passes over the end of the scale so that the direct beam may be seen. It may be necessary to adjust the tilt of the scale by adjusting the three spring loaded screws which are part of the mounting of the scale.

2. The bright diffraction spots should be clearly visible on a wall or screen placed about 2m away. Measure the distance x_0 from the scale to the screen.

3. Determine O, the midpoint between the direct and the specularly reflected (m=0) beams.

4. Measure, with respect to the midpoint, the vertical height y_m of all visible orders.

5. Tabulate your data of y_m^2 and m, the order of the diffracted beam.

ANALYSIS:

1. Plot y_m^2 vs. m on linear paper. Write the equation of the line, as in eq. (6).

2. Calculate y_0 and its uncertainty from the intercept of the line, and compare it to the measured value.

3. From the slope of the line, calculate λ and its uncertainty. Use d = 0.0508 cm (0.02 in).

4. Compare your experimental value of λ to the expected value of 632.8 nm.

EXPERIMENT 8

THE DIFFRACTION GRATING SPECTROMETER

OBJECTIVE:

To study the use of a diffraction grating and to measure the visible spectrum of hydrogen.

REFERENCES:

Fundamentals of Physics by Halliday and Resnick, Geometrical and Physical Optics by Longhurst or Physics by Tipler.

EQUIPMENT:

Spencer spectrometer, diffraction grating, low power lamp, sodium spectral source, hydrogen spectral source.

INTRODUCTION:

When a parallel beam of monochromatic light is incident normally on a transmission diffraction grating, the resulting Fraunhofer diffraction pattern has maxima of intensity at angles θ given by

$$d \sin\theta = m\lambda \quad m = 0, \pm 1, \pm 2, ... \tag{1}$$

where d is the spacing of the rulings, m is the order of the maxima and all angles are measured with respect to the grating normal. The dispersive properties of a diffraction grating can be used to study the spectral composition of a light source in much the same way as was done in THE DISPERSION OF A PRISM experiment. However, good gratings have a much greater dispersive power than do prisms. Consequently they can provide a more detailed picture of the spectra being studied. The dispersion $\mathcal{D}$ of a grating is a measure of the angular separation of adjacent wavelengths and is given by

$$\mathcal{D} \equiv \frac{d\theta}{d\lambda} = \frac{m}{d \cos\theta} \quad m = 0, \pm 1, \pm 2, ... \tag{2}$$

It follows that

1. The angular separation of wavelengths λ and $\lambda+\Delta\lambda$ is proportional to the order m, so the separation of neighboring lines in the 3rd order is at least three times that in the 1st order. (For small angles θ the cosine term is nearly constant).

2. The dispersion is inversely proportional to the grating spacing, that is, it is proportional to the number of rulings per centimeter.

3. Because of the presence of the $\cos\theta$ term in eq. (2), the dispersion in any order will vary throughout the spectrum. For normal incidence the longer wavelengths correspond to larger values of θ, hence the dispersion is greater toward the red end of the spectrum.

PREPARATION:

Before coming to laboratory, calculate and tabulate in the data table, (Table 2) the wavelengths λ_{CALC} of the visible lines in the Balmer series of hydrogen.

PROCEDURE:

1. Look through the telescope at a black piece of paper, and slide the eyepiece in and out until the crosshairs are sharply defined.

2. Sight upon a distant object, and turn the focus ring to bring the object into focus with the crosshairs.

3. Move your eye laterally and make fine adjustments to the focus until there is no relative motion between the 'object' and the crosshairs. At this point, the object image and the crosshairs are in the same plane and have no parallax between them.

4. Align the telescope and the collimator, and view the slit to be illuminated with the low power lamp.

5. Turn the ring on the collimator until the image of the slit is a fine line.

6. Focus this upon the crosshairs by sliding the slit housing in or out, avoiding parallax error as in step (3). Since eq. (1) assumes that the light rays are parallel, proper focussing is important.

7. Rotate the housing without upsetting the focus until the slit is vertical.

8. Clamp the platform, and insert the grating with its rulings vertical into the holder on the spectrometer table.

9. Rotate the spectrometer table until the grating is perpendicular to the collimator.

10. Place the sodium lamp in front of the collimator, and use the telescope to find the zero order (m=0) central maximum. Record the position of the telescope.

12. Similarly locate the 1st order D lines on the left of the central maximum, and record the position and uncertainty of each line.

13. Turn the spectrometer table as necessary until the right and left positions of the 1st order D lines are symmetrically located about the central maximim; it is not important to have perfect symmetry. Clamp the table.

14. Find and observe the 2nd order D lines on the right and left sides. The relative positions of these lines should be the same to within a few degrees. If not, adjust the diffraction grating and repeat step (10) through step (13). The second order D lines should appear in pairs. If they do not appear distinctly as two adjacent, separated lines, reduce the slit width (on the collimator) until they are resolved.

15. Carefully remeasure the left and right positions of the first and second order D lines. If they do not agree with the previous measurements, measure them again to ensure accuracy. Include the uncertainty with each measurement.

16. Replace the sodium lamp with the hydrogen source, and observe the spectrum. **Be sure to turn off the hydrogen lamp when not using it since its lifetime is rather short.** If the spectra appear to be continuous, consult your laboratory instructor who will check and replace it if necessary.

17. Record the left and right positions of each visible line in every possible order in the data table provided.

ANALYSIS:

1. From the data collected with the sodium lamp and eq. (1), calculate d for each D line. Determine the average value of d and its uncertainty, and calculate the number of lines per mm on the grating.

2. Calculate $\Delta\theta/\Delta\lambda$ for the Na D lines and compare it to $\mathcal{D}=m/d\cos\theta$.

3. Calculate your experimentally determined wavelength λ and its uncertainty for each line in the visible spectrum of hydrogen. Compare with the values that you obtained before class.

4. How would the results of the experiment change if it were done under water?

5. Why could you not observe all of the lines in the 3rd and higher orders?

6. Was the angular separation between the red and the blue lines greater for the 1st or for the 2nd order?

TABLE 1: CALIBRATION OF THE DIFFRACTION GRATING

SOURCE: Sodium D lines, $\lambda = 589.00$ nm and 589.59 nm

Wavelength (nm)	Order	Reading Left L (degrees)	Reading Right R (degrees)	L-R (degrees)	$\theta=$(L-R)/2 (degrees)	$\sin\theta$	$d = \dfrac{m\lambda}{\sin\theta}$
589.00	1	........±......	........±......	........±......	........±......	........±......	........±......
589.59	1	........±......	........±......	........±......	........±......	........±......	........±......
589.00	2	........±......	........±......	........±......	........±......	........±......	........±......
589.59	2	........±......	........±......	........±......	........±......	........±......	........±......

Average d= ......±...... nm = ...±...... mm. Number of lines/mm on grating=1/d=......±...... lines/mm

Dispersion of Grating For The Sodium D Lines

$\left(\dfrac{\Delta\theta}{\Delta\lambda}\right)_{\text{1st order}} = $ .... degrees / nm $\left(\dfrac{m}{d\cos\theta}\right)_{\text{1st order}} = $ .... degrees / nm

$\left(\dfrac{\Delta\theta}{\Delta\lambda}\right)_{\text{2nd order}} = $ .... degrees / nm $\left(\dfrac{m}{d\cos\theta}\right)_{\text{2nd order}} = $ .... degrees / nm

TABLE 2: CALCULATION OF WAVELENGTHS USING DIFFRACTION GRATING

SOURCE: ....

Color	Order	Reading Left (L°)	Reading Right (R°)	L°-R°	(L-R)/2=θ	sinθ	λ=d sinθm (nm)	λ_{CALC} (nm)
.....	.....	.....	.....	.....	.....	.....	.....	.....
.....	.....	.....	.....	.....	.....	.....	.....	.....
.....	.....	.....	.....	.....	.....	.....	.....	.....
.....	.....	.....	.....	.....	.....	.....	.....	.....
.....	.....	.....	.....	.....	.....	.....	.....	.....
.....	.....	.....	.....	.....	.....	.....	.....	.....
.....	.....	.....	.....	.....	.....	.....	.....	.....
.....	.....	.....	.....	.....	.....	.....	.....	.....
.....	.....	.....	.....	.....	.....	.....	.....	.....

Visible wavelengths of Hydrogen (λ_{CALC}): _____

EXPERIMENT 9

POLARIZATION OF LIGHT

OBJECTIVE:

To produce various states of polarization of visible light by means of dichroism (selective adsorption), reflection and birefringence.

REFERENCES:

Physics by Tipler, Optics by Hecht and Zajac, Geometrical and Physical Optics by Longhurst, Physics by Halliday and Resnick, and Vibrations and Waves by French.

EQUIPMENT:

Optical bench and accessories, two polarizers, 1/4 and 1/2 wave plates, calcite crystal, photo resistor, d.c. power supply, Weston voltmeter and hook-up wire.

THEORY:

According to Maxwell's theory, light in an isotropic medium is a transverse electromagnetic wave for which the electric and magnetic disturbances are not only perpendicular to the direction of propagation but are also mutually perpendicular to each other. Conventionally, the plane defined by the axis of propagation and the direction of the electric field is taken as the plane of polarization (see fig. 1) so that one may speak of a plane polarized wave or equivalently, of a linearly polarized wave. Usually, however, the light emitted from most common sources is unpolarized. In this case the radiating atoms of the source emit wave trains in which the individual electric field vectors are randomly oriented

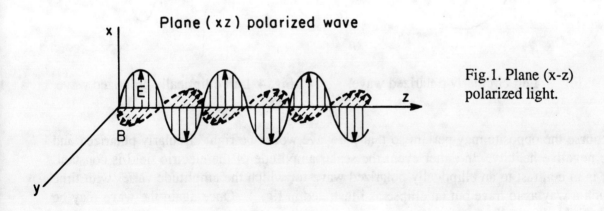

Fig.1. Plane (x-z) polarized light.

space though, of course, they are still perpendicular to the propagation direction. See fig. 2.

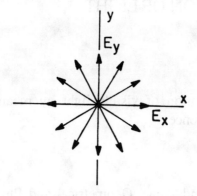

Fig.2. Unpolarized light.

Two other states of polarization are also known to exist, the circular and the elliptical. In a circularly polarized wave the electric vector continuously rotates around the propagation direction. This is shown in fig. 3. When such a wave is viewed head on the electric vector is seen to rotate in an anti-clockwise fashion in which case it is known as a left circularly polarized wave. (In modern terms the wave is said to have positive helicity.)

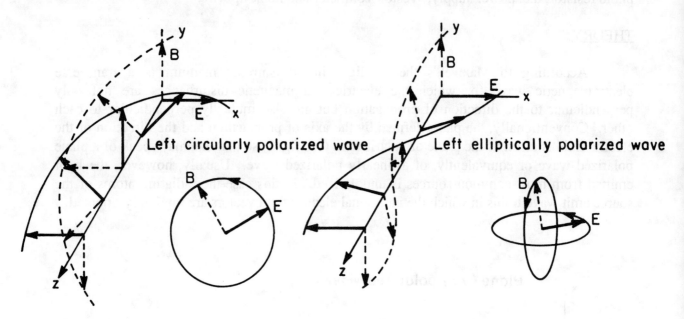

Fig. 3. Left circularly polarized wave.   Fig. 4. Left elliptically polarized wave.

Of course the opposite may pertain so that the wave would be right circularly polarized and have negative helicity. In either event the scalar amplitude of the electric field is constant. This is in contrast to an elliptically polarized wave for which the amplitude varies with time in such a way as to trace out an ellipse as illustrated in fig. 4. Once again the wave may be left or right elliptically polarized.

Any given state of polarization of a light wave may be represented by the superposition of a pair of independent, linearly polarized **component** waves, the electric fields of which are mutually orthogonal.[1] For example, let the component waves have their electric fields oriented along the x and y axes respectively so that

$$E_x = E_{0x} \cos(kz - \omega t)$$
$$E_y = E_{0y} \cos(kz - \omega t + \delta)$$
(1)

where δ is the phase difference between the two. By a suitable choice of the amplitudes E_{0x}, E_{0y} and of δ all four states of polarization may be generated. Since the treatment of elliptical polarization is mathematically somewhat complex, however, it will be omitted here. Fig. 5 illustrates how three common states of polarization are derived from the superposition principle.

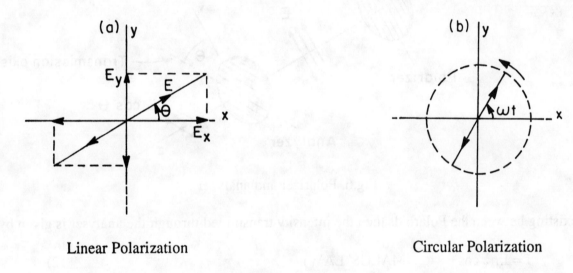

Linear Polarization   Circular Polarization

Fig.5. Superpositions of x- and y- field components.

The principal means by which polarized light is created are those involving dichroism, reflection, birefringence and scattering. Lasers also serve as stable, intense sources of linearly polarized light. Only the first three will be discussed here.

Dichroism refers to the ability of a material to selectively absorb one of the two orthogonal linearly polarized components of an incident unpolarized beam. Such a device is known as a **polarizer** and is itself physically anisotropic. It absorbs one field component while remaining essentially transparent to the other.

---

[1] For a clear discussion of the principle of superposition see French page 29 et seq.

The orientation of the transmitted electric field defines a direction which is known as the **transmission axis** of the polarizer. This will often be referred to in what follows simply as 'the axis'. The well known Polaroid sheet, invented by E. H. Land in 1938, is a very commonly used polarizer. It consists of polymeric material in which the long chain molecules are aligned by stretching during manufacture. Curiously, the transmission axis is perpendicular to the orientation of the molecules. If unpolarized light is incident upon such a Polaroid sheet (the **polarizer**), half the intensity is transmitted, being linearly polarized parallel to the axis, regardless of the orientiation of that axis. This follows because of the complete symmetry of the unpolarized light. Suppose that now a second identical sheet, or **analyser**, whose axis makes an angle θ with the first is introduced as shown in fig. 6. Only the component E cosθ will be transmitted through the analyser. Therefore, if I_{MAX} is the intensity

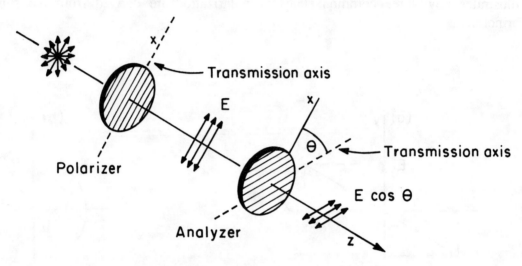

Fig.6. Polarizer and analyzer

existing between the Polaroids then the intensity transmitted through the analyser is given by

$$I = I_{MAX} \cos^2\theta \quad \text{(MALUS' LAW)} \qquad (2)$$

Of course, I_{MAX} is $I_0/2$ where I_0 is the original incident intensity. The significance of the term 'analyser' should now be apparent. An analyser enables one to determine if light is linearly polarized as opposed to being circularly, elliptically or even un-polarized, and in addition the plane of polarization can be found on application of Malus' law.

Reflection of light from an interface between two media, can be either partially or totally linearly polarized, as was first discovered by Sir David Brewster. The degree of polarization depends upon the angle of incidence and upon the indices of refraction of the two media. When the angle of incidence is such that the reflected and refracted rays are 90° to each other the reflected ray is found to be completely polarized. This latter situation is illustrated in fig. 7 where θ_B, known as Brewster's angle, is the particular angle of incidence giving complete polarization of the reflected light..

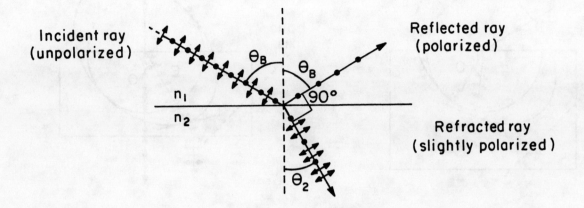

Fig.7. Polarization by reflection.

Now from the geometry, $\theta_2 = 90° - \theta_B$, so upon using Snell's law we may write

$$n_1 \sin\theta_B = n_2 \sin\theta_2 = n_2 \sin(90° - \theta_B) = n_2 \cos\theta_B \qquad (2)$$

or

$$\tan \theta_B = \frac{n_2}{n_1} \quad \text{(BREWSTER'S LAW)} \qquad (3)$$

Note that the electric vector of the reflected light is perpendicular to the plane of the figure so that a sheet of Polarioid with its axis set vertically (sunglasses) would block out all the light (glare) reflected at Brewster's angle from a horizontal surface.

Anisotropy is a characteristic aspect of the crystalline state insofar as optical properties are concerned. A consequence is that the speed of light in such a crystal depends upon the direction of propagation and upon the state of polarization of the light. In the simplest case of (uniaxial) crystals such as quartz, ice, tourmaline and calcite, to name only four, it is found that incident natural light separates into two rays which travel with different speeds and are polarized in mutually perpendicular directions. Such crystals are said to be **birefringent** or **double-refracting**. One of the rays invariably obeys Snell's law and travels with the same speed in all directions so that it has a single, isotropic index of refraction n_O associated with it. Its Huygens' wavelets are spheres. For obvious reasons this ray is known as the ordinary ray (O ray, for short). The other, extraordinary ray (E ray) does not, in general, obey Snell's law nor is its speed the same in all directions. The index of refraction for the E ray is thus found to vary with direction from n_O to, for calcite, a smaller value n_E. The Huygen wavelets are now ellipsoids of revolution, the axis of revolution being a characteristic and unique direction in the crystal known as the **optic axis**. See fig. 8.

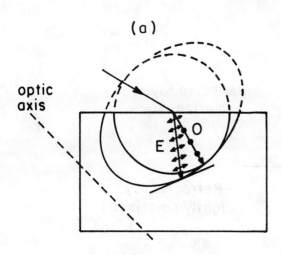 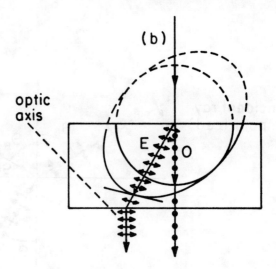

| Snell's law is not obeyed by the E ray. | For normal incidence, double refraction and a speed difference are found. |

Fig. 8. Optic axis at an angle to interface.

The two rays have the same speed (the same index n_O) when propagating along the optic axis and their wave surfaces are tangent there. For the perpendicular direction, the difference in speed is a maximum - the index is n_O for the O ray and n_E for the E ray so that in calcite, for which $n_E < n_O$, the E ray travels faster. The mutually orthogonal states of polarization of the two rays can be easily found in all cases from the following rules:

O ray. The polarization[2] is perpendicular to the plane defined by the O ray propagation direction and the optic axis, and it also is tangent to the (spherical) O wave surface.

E ray. The polarization[2] lies in the plane defined by the E ray and the optic axis and it is also tangent to the (elliposoidal) E wave surface.

These considerations are illustrated in fig. 9 for the simple cases of the optic axis being either parallel or perpendicular to the plane of incidence. (For a more general orientation of the optic axis the situation is very complicated.)

| Optic axis parallel to boundary, parallel to incident plane. | Optic axis perpendicular to boundary, parallel to incident plane. | Optic axis parallel to boundary, perpendicular to incident plane. |

---

[2]For electromagnetic theory applied to material media, the fundamental electric vector is the electric **displacement D** rather than the electric **field E**. The polarization of a wave is therefore more appropriately described by the direction of **D** and it is this direction which is shown in fig. 8. For the O-ray, **D** and **E** are collinear. For the E ray, **D** is tangent to the wavefront in accordance with the rule but **E** is perpendicular to the ray so that **D** and **E**, though co-planar with the optic axis, are not collinear (except for directions along or normal to the optic axis).

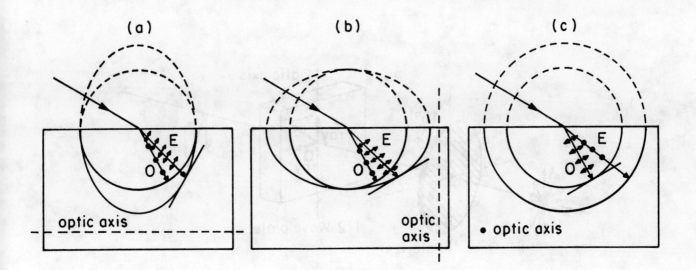

| Snell's law not obeyed by E ray except at normal incidence for which there is no double refraction but a speed difference. | Snell's law not obeyed by E ray except at normal incidence for which there is no double refraction and no speed difference. | Snell's law obeyed by E ray. For normal incidence no double refraction, but a speed difference. |

Fig. 9. Uniaxial negative crystal ($n_E < n_O$).

Birefringent crystals may obviously be used to create linearly polarized beams from natural light. However, they also may be employed to rotate the plane of polarization of linearly polarized light or to generate circularly or elliptically polarized light from linearly polarized light.

1. Suppose that a plate is cut, from calcite say, so that the surface of the plate is parallel to the optic axis as in fig. 9(a) or 9(c). The at normal incidence both the O and E rays will travel the same path but will emerge with a phase difference (because of the different speeds of propagation). If the thickness of the plate is such that the phase difference is π then the plate is known as a 1/2 wave plate. Now let a light ray with linear polarization at 45° to the optic axis enter a 1/2 wave plate. See fig. 10. The emergent ray will have its plane of polarization rotated by $\pi/2$. This happens because the incident polarized ray can be decomposed into two orthogonal components with equal amplitudes - one along the optic

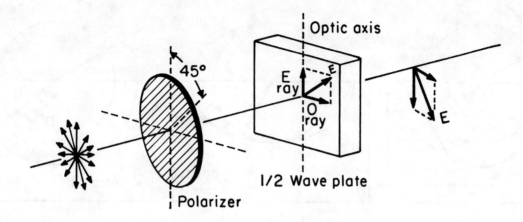

Fig. 10. The 1/2-wave plate.

axis, the E ray, and the other at right angles to it, the O ray. On emergence the components have suffered a relative phase shift of π. Therefore, as reference to fig. 5(b) shows, the resultant electric field is rotated by 90°. Should the angle between the incident polarization and the optic axis not be ± 45°, the O and E. components will be unequal, though the relative phase will remain the same, with the result that the polarization direction will be rotated by less than 90°.

If a 1/4 wave plate is used instead in the above, the emergent rays are $\pi/2$ out of phase so that circularly polarized light is obtained. See fig. 5(c). Should the angle between the plane of polarization of the incident beam and the optic axis not be ± 45°, elliptically polarized light will result.

PROCEDURE:

Transmission Axis

1. Select a polaroid, and mount it vertically on the optical bench.

2. Determine the transmission axis of the polaroid by observing light reflected from the floor (or any other horizontal surface). The reflected light is polarized parallel to the horizontal surface, so when the transmission axis is perpendicular to the horizontal surface, a minimum of light passes through the polaroid. Rotate the polaroid in its frame until a minimum of light is transmitted, and note its angular position. The transmission axis is not necessarily aligned with the 0 or 180° marks. The polaroid will be used at this orientation for the remaining procedures.

O and E Rays

3. Draw a dark dot on a piece of paper, and place the calcite crystal over it. You should see two images of the dot. This is due to the O and E rays pictured in fig. 8(b).

4. Take the second polaroid, use it for an analyser, and observe the images of the dot through the crystal.

5. Rotate the polaroid and notice the images. Record the angle at which one of the images just disappears. Continue to rotate the polaroid, and record the angle at which the other dot just disappears.

6. Rotate the crystal in the horizontal plane, and describe what you observe.

Verification of Malus' Law.

7. Mount the photoresistor, the polarizer, and the lamp on the optical bench about 30 cm apart, and adjust their heights until the photoresistor is centered in the light beam passing through the polarizer. The photoresistor is a device with a conductance that depends linearly on the intensity of light which falls on it.

8. Set up the circuit shown in fig.11.

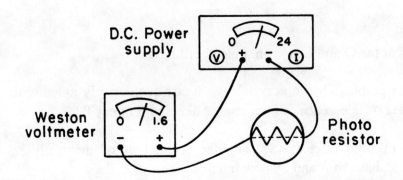

Fig. 11. Circuit used to verify Malus' Law

The voltmeter is one of low resistance, but in this circuit it acts as a high resistance ($\approx 160\Omega$) ammeter. The voltmeter will be used to measure (up to a constant) the light intensity incident on the photoresistor. It should be set on the 1.6V scale, and the power supply should be adjusted to give about full scale deflection.

9. Place the analyser on the optical bench close to the photoresistor, and record the voltmeter reading for various angular orientations of the analyser. Take readings every 10 degrees for one complete revolution. Around minima and maxima, take additional readings. Plot the voltmeter reading vs. angular displacement as you proceed.

degrees for one complete revolution. Around minima and maxima, take additional readings. Plot the voltmeter reading vs. angular displacement as you proceed.

Rotation of the Plane of Polarization.

10. Align the transmission axis of the analyser with the axis of the polarizer. The voltmeter should read a maximum at this point.

11. Place the 1/2 wave plate between the two polaroids, and orient the optic axis of the plate so that it is at 45° with respect to the transmission axis of the polarizer. Note that the voltmeter reads a minimum.

12. Record the voltmeter reading for a 180° turn of the analyser. Plot the voltmeter reading vs. angular displacement on the same graph used in step (9) as you proceed.

Circularly Polarized Light.

13. Remove the 1/2 wave plate and replace it by the 1/4 wave plate. Orient the optic axis of the plate so that it is at 45° with respect to the transmission axis of the polarizer.

14. Record the voltmeter reading as a function of the angular orientation for a full revolution of the analyser, and make a plot on the same graph before.

ANALYSIS:

1. Verify that the polarizations of the O and E rays are indeed orthogonal.

2. Does the shape of your first graph (when no crystal was used) qualitatively agree with Malus' Law as expressed in eq. (2)? In essence, does it appear like a plot of $\cos^2\theta$?

3. Compare the three plots of light intensity vs. angular orientation of the analyser. Describe which plot undergoes a phase shift and by how much.

EXPERIMENT 10

HOLOGRAPHY

OBJECTIVE:

To study holography and to make a hologram.

REFERENCES:

Optics by Hecht and Zajac, Exploring Laser Light by Kallard or Introduction to Modern Optics by Fowles.

EQUIPMENT:

He-Ne laser, two laboratory tables, stone slabs, one 1-meter and one 1/2 meter optical bench, component table, 3" x 5" plane front surface mirror, 8 mm focal length converging lens, lens holder, 3 mm aperature in large shield with holder, film holder on magnetic mount, SO-253 film, black cloth, light meter.

HOLOGRAPHY:

In converntional black and white photography, the light scattered or reflected from and object is collected by the camera lens and imaged upon the film. The information that is so recorded is, of course, merely a mapping of the intensity distribution over the surface of the object. The resultant photograph, while it is certainly quite a good representation of the object, is nontheless purely two-dimensional. This deficiency can be ascribed to the fact that no information about the phase of the light wave emanating from the object was recorded on the film. If both the phase and the amplitude of the original wave from the object could be recorded in some fashion, as in a hologram, then under suitable illumination of the hologram, the resulting light field would be indistinguishable from the original one. Hence one could view a reconstructed, three dimensional image of the object, just as if the latter were actually present. Indeed as one views such a hologram from different angles, different perspectives of the image are seen! The early work in this field was done by Dennis Gabor in the later forties but it was destined to remain in near oblivion for fifteen years until, with the advent of intense coherent light sources (lasers), there was a resurgence of interest during the mid sixties. Gabor received the Nobel prize in 1971 for his pioneering work in holography.

Gabor's basic idea is illustrated, in modern form, in fig. 1. A broad beam of monochromatic coherent light is partially intercepted by a mirror and reflected toward the photographic film in the holder. This portion is known as the **reference wave**. The remainder of the incident beam impinges upon the object and some of it is scattered back towards the film holder. This scattered **object wave** is very complex since its amplitude and

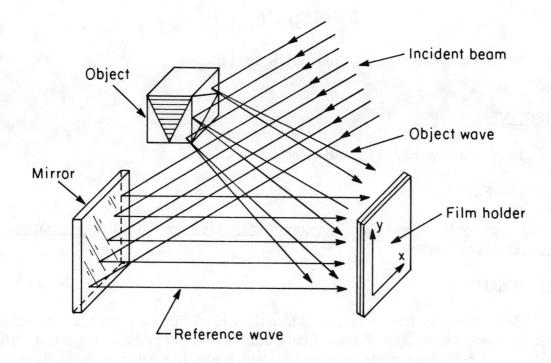

Fig. 1. Basic geometry of film exposure.

phase vary in a complicated way (determined by the structure of the object) across the wavefront. At the film holder the coherent reference and object waves interfere to produce a stationary and intricate interference pattern which is recorded on the film. The developed film is known as a **hologram**. In ordinary light there is no visible recognizable image as such. Rather, under a microscope one would merely see an interference pattern of considerable complexity. But when a parallel beam of light passes through the developed film, images which resemble the original object as viewed from any direction can be seen. These arise through diffraction of the incident beam by the complicated interference pattern on the developed film.

The theory for this is given below, in the Appendix. Our discussion at this point is more qualitative, but can be completely justified through the theory. The object can be viewed, for optical purposes, as being made up of a collection of spherically outgoing Huygens wavelets reflected from each point of the object. One can consider each point independently, then superpose to obtain a final result. The light from one of the points, O, and the reference beam are shown in fig. 2, together with a schematic depiction of the coherent superposition (intensity) as it exposes the photographic film. Wherever the angular separation of incident rays is large, the interference pattern has a small spatial structure; and visa versa, as shown in the figure. Upon exposure and development this spatially dependent interference pattern is permanently recorded. In fig. 3 one sees the result of allowing the reconstruction beam C to pass through such a grating. Where the grating spacings are small, the diffraction angles are large, and visa versa. The first order diffraction rays, to either side

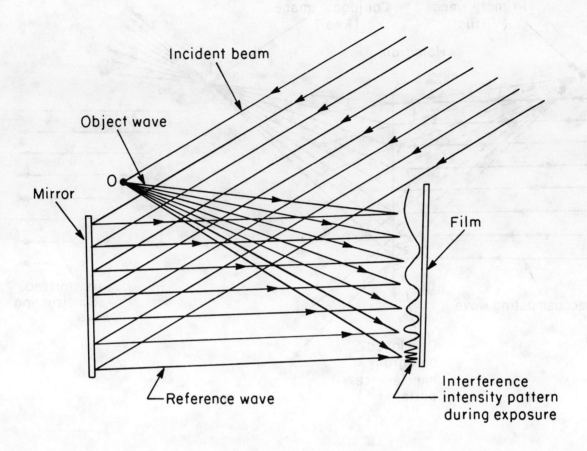

Fig. 2. Film exposure for a point object, O.

of the transmitted reconstruction beam direction, are shown. As seen in the figure some of the rays appear to be coming from the original object position, now labelled I, and others converge to, then diverge from, a conjugate point I'. The rays tracing back to I represent the effect of a virtual image, optically identical, in all respects, to the original object. The rays converging, then re-emerging from I' represent a real image.

By replacing the point object, O, by an extended object, one would obtain, at I an extended image with exactly the same optical properties as the original object (including the three dimensional appearance) and at I' an image which will be a front-to-back inversion of the original object. Note: for the holograms which you are making, the virtual image I will be difficult to view.

Do not look toward the laser in an attempt to see the virtual image.

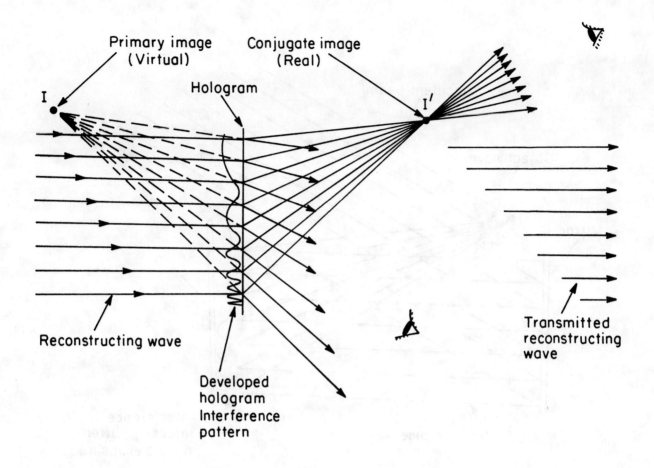

Fig. 3. The virtual and real image, I and I' of original object O.

PREPARATION:

Bring to class a suitable object from which a hologram can be made. Ideal objects are those which reflect light with specular highlights and contrasts, are 3-dimensional, rigid, and are no larger than approximately 3cm x 3cm x 3cm. Alternatively several small objects may be arranged to form a scene.

PROCEDURE:

1. Turn on the laser so it will be stable during the experiment.

2. Arrange the apparatus as shown in fig.4. It is important that the distances OF and MF differ by less than the coherence length of the laser (~2 to 3 in). Position the object so that its 3-dimensional aspects are readily apparent from various points in the plane of the film holder. Make sure that the object is fully illuminated near the center of the incident light beam (or slightly into the wing), and that the mirror is illuminated by the farther wing. The optimum ratio of object to reference beam brightness is 1:3. This will only occur for bright

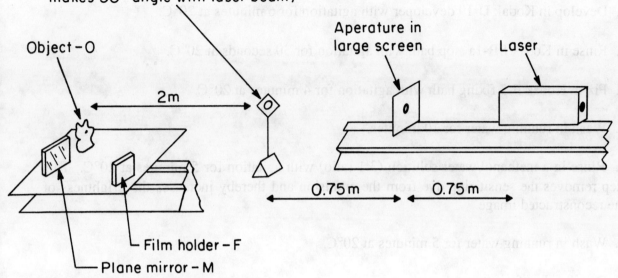

Fig. 4. Exposure set-up.

objects and a much higher illumination of the object than the mirror. However, even a ratio of 1:10 is satisfactory. Adjust the plane mirror M_2 so the reference beam uniformly illuminates the film holder. The 30° tilt of the 8mm lens ensures that a broad wing of the laser light will illuminate the mirror evenly.

3. With the aid of your instructor, measure the intensity of the incident beam at the object position.

4. Record your instructor's recommendations for film exposure time. Use the Metrologic photometer on the 0.003 range, and connect it to the internally illuminated microammeter. Remove the hood from the sensing element before measuring the intensity. A reading of 5μA indicates an exposure time of approximately 8 sec; 10 μA corresponds to 4 sec, etc.

5. Using the storage can and an exposed piece of film, practice removing the film, placing it in the film holder, making the exposure, and returning the film to the can. Eventually you will need to do this in the dark.

Exposure.

6. The instructor will direct the class in making the exposures. One student in your lab group should handle the film while another student controls the shutter, a black piece of cloth placed in the path of the laser beam near the laser. Be sure that no extraneous light can reach the film holder. During the actual exposure, the table must not move; therefore, everyone must sit completely still, and all unnecessary movement must cease.

Film Processing.

1. Develop in Kodak D-19 developer with agitation for 6 minutes at 20°C.

2. Rinse in Kodak SB-1a stop bath with agitation for 30 seconds at 20°C.

3. Fix in Kodak F-5 fixing bath with agitation for 4 minutes at 20°C.

4. Wash in running water for 10 minutes.

5. Rinse in a methanol/water solution (3:1 ratio) with agitation for 5 minutes at 20°C. This step removes the sensitizing dye from the emulsion and thereby increases the brightness of the reconstructed image.

6. Wash in running water for 5 minutes at 20°C.

7. Dry in a dust free atmosphere.

If time is short, washing in step (4) can be shortened, and steps (5) through (7) can be omitted.

Reconstruction
8. Examine the hologram and notice the presence of a large scale ripple, rings, etc. These are not the microscopic interference pattern characteristics of the object, but are diffraction patterns produced by dust, flaws, and scratches on the optical components.

9. To view the real image I', place the hologram in a laser beam and place a screen beyong the hologram by a distance equal to the original object to film distance (about 20cm). Move the hologram until you get an optimal image on the screen.

> Do not look into the laser as your attempt to locate this image.

APPENDIX. THEORY OF HOLOGRAPHY

Exposure. Let the reference wave over the (xy) plane of the film be written in the form

$$E_R(x,y) = E_{0R}\cos[\omega t + \phi_R(x,y)] \tag{1}$$

where E_{0R} is the (constant) amplitude and $\phi_R(x,y)$ describes how the phase of the reference wave varies over the film. (Actually $\phi_R(x,y)$ is determined completely by the geometrical disposition of the film plane relative to the mirror and is therefore some fixed function). The object wave may be similarly expressed as

$$E_O(x,y) = E_{0O}(x,y) \cos[\omega t + \phi_O(x,y)] \tag{2}$$

where now the amplitude $E_{0O}(x,y)$ as well as the phase $\phi_O(x,y)$ are generally both very complicated functions of position, depending as they do upon the nature of the object. These two waves superpose and produce an interference pattern on the film. The resulting intensity $I(x,y)$ is, apart from a constant, given by the mean value of $(E_R + E_O)^2$, thus

$$I(x,y) \propto E_{0R}^2/2 + E_{0O}^2(x,y)/2 + E_{0R} E_{0O}(x,y)\cos[\phi_R - \phi_O(x,y)] \quad . \tag{3}$$

Note that the position of the intensity maxima and minima are determined largely by the relative phase of the reference and object waves. Let us assume that $E_{0O} \ll E_{0R}$ so that

$$I(x,y) \cong E_{0R}^2/2 + E_{0R} E_{0O}(x,y) \cos[\phi_R - \phi_O(x,y)] \quad . \tag{4}$$

Because of the presence of $E_{0O}(x,y)$ and $\phi_O(x,y)$ in this equation the hologram constitutes a complete record of the amplitude and phase of the wave emanating from the object.

Reconstruction. Let the reconstruction wave E_C fall on the developed hologram at the same angle as did the reference wave during exposure. Then the phase of E_C across the (x,y) plane of the hologram will merely be the $(\phi_R(x,y)$ discussed above, therefore

$$E_C(x,y) = E_{0C} \cos[\omega t + \phi_R(x,y)] \quad . \tag{5}$$

Under these circumstances the wave E_T transmitted through the hologram will be proportional to the product $E_C(1-KI(x,y))$, where K (assumed $\ll 1$) depends upon the exposure and processing of the film. It is readily seen[1] that the transmitted wave is

$$E_T(x,y) = E_{0C}\left(1 - \frac{1}{2}KE_{0R}^2\right)\cos[\omega t + \phi_R(x,y)]$$

$$- \left(\frac{1}{2}K E_{0C} E_{0R}\right) E_{0O}(x,y) \cos[\omega t + 2\phi_R(x,y) - \phi_O(x,y)] \tag{6}$$

$$- \left(\frac{1}{2}K E_{0C} E_{0R}\right) E_{0O}(x,y) \cos[\omega t + \phi_O(x,y)] \quad .$$

---

[1] Use the trigonometric identity: $\cos x + \cos y = 2 \cos\frac{1}{2}(x + y) \cos\frac{1}{2}(x - y)$.

Of the three terms representing the light coming from the hologram consider the last one. Apart from the constant factor of $-\frac{1}{2}KE_{0C}E_{0R}$ it exactly the divergent scattered wave that originally emanated from the object - compare eq. (2). So indeed one can reconstruct the original (very complicated) wave, from the hologram, as if the original object itself were present.

In eq. (6), the second term represents the convergent wave which eventually reemerges from the real or **conjugate** image. The leading term in eq. (6) is merely the transmitted reconstructing wave, amplitude modulated to be sure, but containing no phase information concerning the object.

EXPERIMENT 11

CHARGE TO MASS RATIO OF THE ELECTRON

OBJECTIVE:

To measure the charge to mass ratio (e/m) of the electron by the method of magnetic deflection.

REFERENCES:

Physics by Halliday and Resnick, Physics by Tipler, Physics by Serway.

EQUIPMENT:

Leybold-Heraeus beam tube with Helmholtz coils and measuring device; Harrison 6284 A d.c. power supply; DMM; Weston d.c. ammeter; Central Scientific 79552 power supply, 30cm scale; hook-up leads.

INTRODUCTION:

Imagine that an electron of charge e and velocity $\mathbf{v}$ enters a region in which there exists a uniform magnetic induction field of magnitude $\mathbf{B}$. The magnetic force (Lorentz force) acting on the electron is determined according to

$$\mathbf{F} = e\, \mathbf{v} \times \mathbf{B} \tag{1}$$

The force is normal to the velocity; therefore, it can do no work on the electron, and the speed of the electron will remain the same. However, the force will alter the direction of the electron. For the case of $\mathbf{B}$ perpendicular to $\mathbf{v}$, the electron travels in a circular motion of radius r and with a centripetal force supplied by the magnetic field. Therefore, the Lorentz force equals the centripetal force on the electron,

$$evB = \frac{mv^2}{r}. \tag{2}$$

Solving for the charge to mass ratio,

$$\frac{e}{m} = \frac{v}{Br}. \tag{3}$$

If the electron is accelerated from rest by an accelerating voltage V, then its kinetic energy is given by

$$\frac{1}{2}mv^2 = eV .\qquad(4)$$

Solving for v, yields

$$v = \sqrt{\frac{2eV}{m}} .\qquad(5)$$

Substituting eq.(5) into eq.(3) gives

$$\frac{e}{m} = \frac{2V}{B^2 r^2} .\qquad(6)$$

The apparatus used in this experiment to provide a uniform magnetic field is a pair of Helmholtz coils, as shown in fig.1.

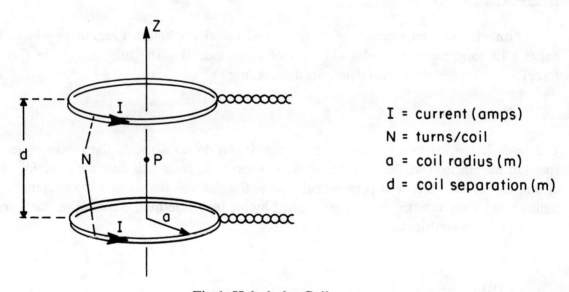

I = current (amps)
N = turns/coil
a = coil radius (m)
d = coil separation (m)

Fig.1. Helmholtz Coils.

If the coil separation is made equal to the coil radius, then the magnetic field at or near the midpoint P on the z-axis is in the z-direction and is given by

$$\mathbf{B} = \frac{8\mu_o NI}{5^{3/2} a} \mathbf{k} \qquad(7)$$

133

where $\mu_0 = 4\pi \times 10^{-7}$ weber/amp·m.

The electron beam is formed inside a glass sphere which contains a residual pressure of approximately 10^{-2} torr of nitrogen. Collisions of the electrons with the nitrogen molecules cause the latter to emit a faint bluish radiation along the track of the beam which can be visible in a dark room. The beam itself is produced by the electrode arrangement shown in fig.2.

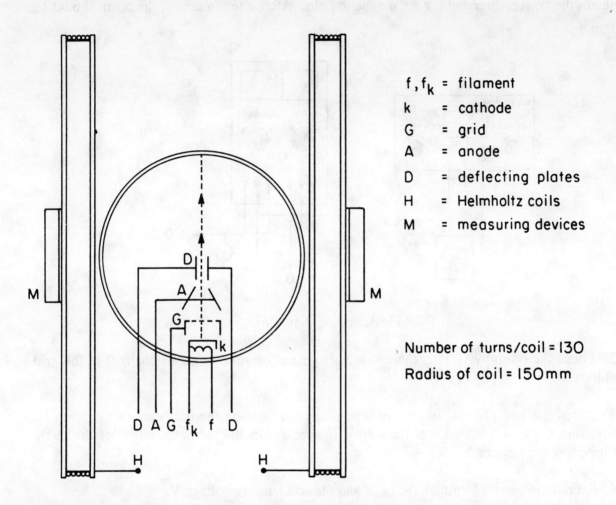

Fig.2. Electrodes which produce the electron beam, and Helmholtz coils.

Electrons are emitted from the indirectly heated cathode K, focused by the Wehnelt cylinder (or grid) G, and accelerated by the conical anode A. The electrostatic deflecting plates D are not used in this experiment and should both be kept at the same potential as the anode. The Helmholtz coils which provide the magnetic field are also shown in fig.2. The Helmholtz coils have a 15 cm radius, are spaced by 15 cm and have 130 turns per coil. A device M consisting of two parts—one with a mirror and the other with two riders for measuring the diameter of the circular electron beam—is attached to the coils.

PROCEDURE:

1. Set up the apparatus with the wiring as illustrated in fig. 3. Set the filament voltage at 6.3V a.c., the grid G at about -10V with respect to the cathode (measure this with the voltmeter on the power supply), and the anode voltage between 150 and 300 volts positive with respect to the cathode (measure this with the DMM). Be sure that the deflection plates are connected to the anode. The maximum current through the Helmholtz coils is 2 amps; do not exceed this value! It may be necessary to interchange the leads to the coils in order to obtain the correct direction for the magnetic field. After a few minutes, the beam should be visible.

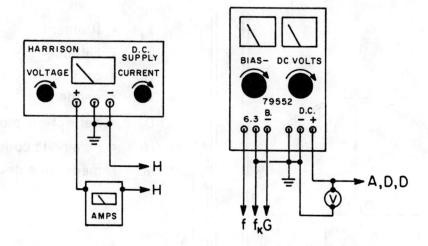

Fig.3. Wiring diagrams.

2. Focus the beam by varying the anode voltage and by making small changes in the grid voltage.

3. When the best possible focus has been obtained, measure the diameter of the beam by adjusting the two riders so that the mirror images are in line with the edges of the circle formed by the electron beam.

4. Record the current I through the coils and the accelerating voltage V.

5. Repeat steps (1) through (4) for various several values of current.

6. Tabulate your values of V, I, and r.

ANALYSIS:

1. Calculate the magnetic field B for each current.

2. Using eq. (6), calculate the charge to mass ratio of the electron for each value of B.

3. Average your results for e/m, and use the deviation from the mean as an estimate of your uncertainty.

4. Compare your average result for e/m to the accepted value of 1.759×10^{11} coul/kg. Do the values agree to within your estimated uncertainty?

5. From your experimental determination of e/m, calculate the mass of the electron m, and compare it to the accepted value of 9.11×10^{-31} kg. The value of *e* is given in Experiment 12, Millikan's Oil Drop Experiment.

EXPERIMENT 12

MILLIKAN'S OIL DROP EXPERIMENT

OBJECTIVE:

To measure the fundamental quantum of electric charge (e) by observing the behavior of charged droplets of oil.

REFERENCES:

Foundations of Modern Physics or Physics by Tipler and University Physics by Sears and Zemansky.

EQUIPMENT:

Millikan's oil drop apparatus complete with microscope and atomizer, Central Scientific 79552 power supply, Keithly 191 digital multimeter, digital stop watch, 15 cm scale in holder, hook up leads and 6" piece of #28 wire.

INTRODUCTION:

Consider a spherical oil droplet of mass m and radius r falling freely in air under the force of gravity. According to Stokes[1] it will reach a constant terminal speed v_g given by

$$mg = 6\pi r \eta \, v_g \qquad (1)$$

where g is the acceleration due to gravity and η is the coefficient of viscosity of air. If the droplet also has a net charge e_n (= ne where e is the quantum of charge assumed to be positive and n is an integer), then if an upwardly directed electric field E of sufficient magnitude is applied, the net upward force acting on the droplet will be $e_n E - mg$ and so its upward terminal speed v_f will be given by

$$e_n E - mg = 6\pi r \eta v_f \, . \qquad (2)$$

On combining eqs. (1) and (2) to eliminate mg, one obtains

---

[1] If a sphere, subject to an external force F, moves in a homogeneous medium, a viscous force proportional to its instantaneous speed opposes to the motion. When the viscous force becomes equal and opposite to F no further acceleration is possible and the sphere thereafter travels with a constant, or terminal, speed. This is known as Stokes' Law after Sir George Stokes who first enunciated it in 1845. The constant of proportionality is $6\pi r \eta$.

$$e_n = \frac{6\pi r \eta}{E}(v_g + v_f). \tag{3}$$

Since the mass of the droplet is $\frac{4}{3}\pi r^3 \rho$, where ρ is the density of the oil from which the droplet is made then from eq. (1) an expression for the radius of the droplet follows,

$$r = \left(\frac{9\eta v_g}{2\rho g}\right)^{1/2}. \tag{4}$$

Substitution of this result into eq. (3) yields

$$ne \equiv e_n = \frac{4\pi}{3}\left(\frac{9\eta}{2}\right)^{3/2}\left(\frac{v_g}{\rho g}\right)^{1/2}\frac{(v_g + v_f)}{E}. \tag{5}$$

Clearly, measurement of the two terminal speeds v_g and v_f coupled with a knowledge of the electrostatic field E enables the charge e_n of a droplet to be found. If this is repeated for a number of droplets each having, in general, a different net charge, then a series of values $...e_{n_i}...$ will be obtained. Examination of this series should reveal a lowest common divisor (e) and, for each member of the series, and an associated integer n_i.

The analysis just given is the essence of Robert A. Millikan's classic "oil drop experiment" first performed in 1909 and repeated on thousands of droplets. He obtained for e the value of 1.591×10^{-19} coul, thereby demonstrating the quantization of charge. The presently accepted value for e is 1.602×10^{-19} coul, the small discrepancy being ultimately traced to Millikan's use of an incorrect value for the viscosity of air. In Millikan's very precise work it was found necessary to make two corrections to the theory given. These will not be used in the present experiment but are mentioned for the sake of interest.

(i) The effective mass of a droplet is $\frac{4}{3}\pi r^3 (\rho-\sigma)$, where σ is the density of air, thus correcting for the buoyancy effect of air; and

(ii) Stokes' law is amended to allow for inhomogeneities in the air, so that the right hand side of eqs. (1) and (2) should be written in the form $6\pi r\eta v(1+(b/pr))^{-1}$. Here, b is a constant and p is the pressure of the air, r being the radius of the droplet as before.

In essence the oil drop apparatus consists of two parallel conducting plates, P, separated by a distance, d, as shown in fig. 1. Oil drops are sprayed in through a small hole in the top plate and are illuminated by a lamp, L. A 15 power microscope, M, with a ruled graticule in the eyepiece, is used to observe the drops. An electric field E can be created in the gap between the plates by applying a potential difference ΔV to them (E = $\Delta V/d$). A reversing switch, R, enables one to either change the direction of the electric field or to short

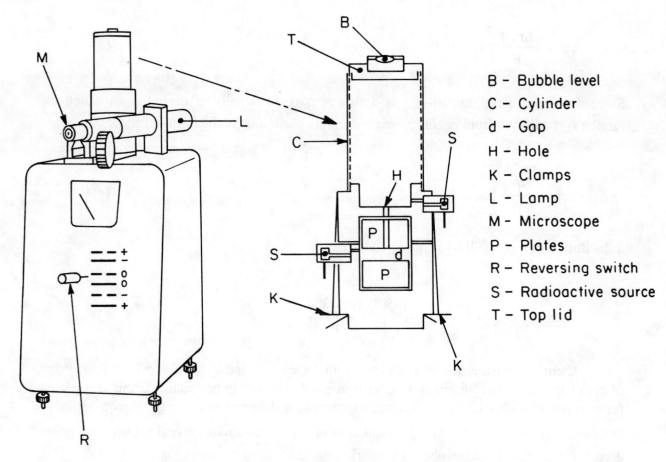

Fig.1. CENCO #71263 experimental apparatus.

the plates together so that no field exists. Two radioactive sources, S, (short range α particle sources) are also provided-- the upper one assists in charging the oil drops before they enter the gap while the lower one can be used to alter the charge on a drop when it is between the plates. These sources are "switched on" by turning the indicating lever to the vertically upwards position.

PROCEDURE:

1. Connect the power supply and the DMM (on the 10^3 range) across the appropriate terminals at the rear of the oil drop cabinet.

2. Carefully level the cabinet by adjusting the four thumb screws in its base and observing the bubble level B in the top of the lid T of the chamber.

3. Remove the aluminum cylinder C, and insert a fine wire through the hole in the top plate. Focus the microscope on the wire so that the drops used later in the experiment will be clearly visible.

4. Accurately set the graticule (the scale in the focal plane of the microscope) to the vertical wire. There should be no parallax between the image of the wire and the graticule.

5. Remove the top plate assembly by releasing the two clamps K, and insert the millimeter scale in its holder. Adjust the position of the scale until it is in focus, and calibrate the graticule.

6. Reassemble the chamber except for the top lid.

7. Prepare the atomizer by holding the tip against a piece of paper until a mist emerges. With the tip of the atomizer turned downward, spray a fine mist of oil into the chamber.

8. Replace the top lid and view the drops through the microscope. Note that the microscope produces an inverted image, so the drops appear to "fall upwards."

9. Select a drop that takes about 10 seconds to travel upward between two of the longer marks on the graticule, and try to reverse its direction by using the reversing switch to apply a positive or negative potential. Use a potential of 100 to 200 V, and adjust it as is necessary. Remember that the oil drops can have positive, negative, or zero charge. If it is seen that the motion of the drop is not parallel to the graticule, then the cabinet has not been levelled properly.

10. Under zero field, time how long it takes the drop to rise a certain number of scale divisions. Record the time t_g and the distance d_g. The observer should start and stop the watch while the lab partner records the data. Sometimes the drop might suddenly speed up or slow down. If this happens, the drop has changed its charge (due to a cosmic ray for example) and should be treated as a new drop.

11. Without allowing the drop to leave the field of view, reverse its direction by applying a field. When the drop is approximately at its starting point, reverse the field such that the drop rises under the influence of both gravity and the electric field. Record the potential V.

12. Time how long it takes for the drop to fall a certain number of scale divisions, and record the time t_f and the distance d_f.

13. Localize another drop, and repeat steps (10) through (12). If you need to spray another mist into the chamber, you may do so, but it should not be necessary for each trial.

14. Record and tabulate the data for 5 drops.

ANALYSIS:

1. Calculate v_g for each drop.

2. Calculate v_f for each drop.

3. Given that the plate separation d is 3mm, calculate E for each drop. Remember that for parallel plates, $V = -Ed$.

4. Given that the acceleration due to gravity in State College is g=9.8012 m/sec², the viscosity of air at 18°C is η=1.83 x 10⁻⁵ kg/m·sec, and the oil density for Octoil S diffusion pump oil is ρ=909 kg/m³, calculate e_n for each drop.

5. Determine n and the least common denominator e for each value of e_n. Average your values of e and compare to the accepted value.

6. Calculate the deviation from the mean, and use this for an estimate of your uncertainty. Does your experimental value equal the accepted value within the estimated uncertainty?

EXPERIMENT 13

THE FRANCK-HERTZ EXPERIMENT

OBJECTIVE:

To demonstrate the quantum nature of the atom and to measure the energy of its first excited state.

REFERENCES:

Foundations of Modern Physics by Tipler, Experiments in Modern Physics by Melissinos or Concepts of Modern Physics by Beiser.

EQUIPMENT:

Franck-Hertz tube, copper-constantan thermocouple, oven, 0-110 volt a.c. power source and Weston a.c. voltmeter, two Keithley 191 digital multimeters, Harrison 6284A d.c. power supply, Central Scientific 79552 power supply, Keithley 610B electrometer, 1.5 volt dry cell, 10K:100K voltage divider, Hewlett Packard 7035B x–y recorder and hook–up leads.

INTRODUCTION:

In 1914 James Franck and Gustav Hertz devised a simple experiment which verified the quantum nature of atoms–as had been proposed a year earlier by Neils Bohr.[1] The heart of the apparatus in its modern form consists of an evacuated glass tube containing a drop of mercury, a filament f, an indirectly heated cathode K, a space charge grid g_1 (the purpose of this will be discussed later under PROCEDURE), an accelerating grid g_2, a collector A and a shield S, as shown schematically in fig. 1.

Electrons are emitted from the cathode and accelerated by a variable voltage V_2 applied to the grid g_2, thereby acquiring kinetic energy $(1/2) mv^2 = eV_2$ when they arrive at the vicinity of g_2. The tube meanwhile is heated to $\approx 180°C$ so that the space is filled with mercury atoms in the vapor phase, the pressure being ≈ 9 torr and the mean free path for electrons being $\approx 5 \times 10^{-5}$ m. As V_2 is slowly increased the electrons acquire kinetic energy and undergo **elastic** collisions with the mercury atoms but lose very little energy in doing so (why?) and are able to reach the collector A. During this stage the observed current increases smoothly with increases in V_2. However, at a particular value of V_2 the current suddenly drops! Now the collisions are **inelastic**, the electrons emerge with very little energy and are unable to reach the collector A which is maintained at a repulsive potential of

---

[1] Apparently neither Franck nor Hertz had heard of Bohr's theory when they did the experiment.

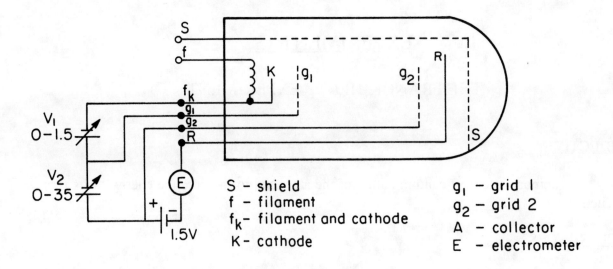

Fig.1. The Franck-Hertz tube.

1.5 volts with respect to g_2. The participating mercury atoms on the other hand have each absorbed the electron's energy and have made a quantum transition from the ground state ($6s6s^1S_0$) to the first excited state ($6s6p^3P_1$) where they remain for a short time ($\approx 10^{-8}$s) before decaying back to the ground state with emission of ultraviolet photons of wavelength 253.7 nm corresponding to an energy of 4.88 eV.

The current rises once again as V_2 is further increased, inelastic collisions occur, of course, but now the electron has sufficient energy left over to enable it to reach the collector. At $V_2 \approx 9.8$V the current drops once more. This corresponds to the electron undergoing <u>two</u> inelastic collisions with two different mercury atoms and emerging with little energy. The process repeats as V_2 is increased and as many as seven peaks may be obtained by the careful experimenter before the mercury vapor ionizes and a glow discharge sets in. The separation between the peaks (see fig. 2) should be 4.88V.

It is remarkable that the quantum nature of an atom may be observed using little more than a common voltmeter and ammeter. In principle, it is possible to observe other peaks, spaced at different intervals, corresponding to excitation of the mercury atoms to levels higher than the first. This possibility depends upon the detailed variation of the potential within the tube, but in the usual setup they are not seen.

<u>PROCEDURE</u>:

1. The success of this experiment depends critically upon the temperature of the tube. If the density of the vapor is low, the electron current will be large (fewer collisions) and the dips will be less pronounced. Too large a density gives rise to very small currents that are difficult to measure. A temperature of 180°C is recommended and may be monitored by the copper-constantan thermocouple, the sensing junction of which is inserted into the oven wall from the rear, the other ends being connected to a digital multimeter on the 200 mV range.

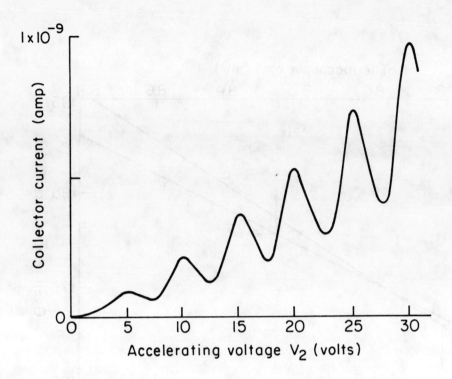

Fig. 2. Collector current vs. V_2

("Null" the multimeter before use.) From fig. 3 obtain the emf in mV corresponding to the existing room temperature, graph (a) with the lower emf scale, and add this to the observed reading from the multimeter. The actual oven temperature may then be read from graph (b) using the upper emf scale and the right hand temperature scale.

2. At the beginning insert the Franck-Hertz tube into the oven and apply 110 volts a.c. for exactly 8 minutes and then reduce to 48 volts; after a total of ≈20 minutes has elapsed, minor adjustments should be made as indicated by the thermocouple. Note that the oven heater leads are blue and black—the yellow/green one is a ground lead. The oven power used is either a Variac or a Cenco power supply in conjunction with the Weston a.c. voltmeter (use the appropriate ranges on this instrument).

3. While the oven is being stabilized, the remainder of the apparatus should be wired up as shown in fig. 4. Be careful to observe the correct polarities of all d.c. voltage sources. Because the <u>minimum</u> d.c. voltage obtainable from the 79552 power supply is ≈13 volts (rather than 0) the smallest voltage that can be applied to g_2 is 1/11 of this (why?). The meter (2.4 V range) on the Harrison d.c. power supply should be used to measure the voltage applied to g_1. This will lie between 0 and 4 volts and is necessary to overcome space charge

> Note: Since individual ovens and tubes vary somewhat, the numbers given for oven voltages and for the electrometer range setting are suggested values only.

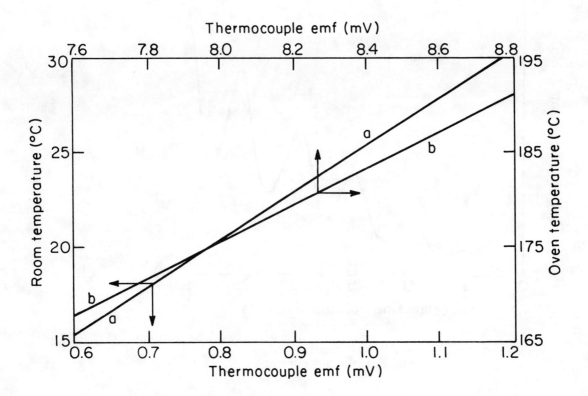

Fig. 3. Thermocouple emf's for room temperature (a) and oven temperature (b).

effects in the tube. The best value may be found by trial and error. The electrometer should be set on either the -3×10^{-9} or -3×10^{-8} amp range. (Why are negative ranges chosen?)

4. When the circuits have been checked by the instructor and the oven temperature is at or near 180°C, slowly turn the knob on the 79552 power supply clockwise thereby increasing the voltage V_2. Several dips in the electrometer current should be observed as you do so. An upper limit to V_2 is ≈30V–if this is exceeded a gas discharge will occur in the tube and the electrometer needle will fly off scale. If this event should occur at any time, IMMEDIATELY reduce V_2 to its minimum value. Reduce V_1 by 0.2 volt and repeat the above; in this way, an optimum value for V_1 may be found, such that no discharge will occur for $V_2 \leq 30V$.

5. The laboratory instructor will now connect an x–y recorder to the apparatus. The 3 volt output at the rear of the electrometer will be connected to the y axis while V_2 will be connected to the x axis. A graph resembling fig. 2 will then be obtained. Calibration marks for the x axis will also be provided so that the voltage separation between adjacent peaks may be measured from the graph. The average value should be computed and compared with 4.88 eV as determined by spectroscopic methods.

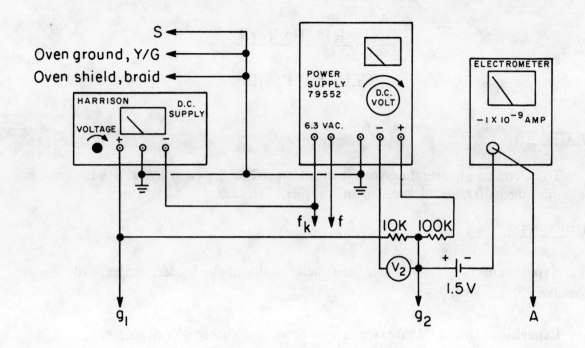

Fig. 4. Circuitry.

ANALYSIS:

1. Measure the voltage separations between adjacent peaks of the chart recording.

2. Determine the average separation, and compare with 4.88eV as inferred from spectroscopy.

EXPERIMENT 14

NUCLEAR PHYSICS

OBJECTIVE:

To determine the operating voltage of a Geiger-Mueller counter and use the counter to determine the half-value thickness of an absorber.

EQUIPMENT:

Experiment (a): G–M tube, tube stand and sample holder, scaler with timer, radioactive ^{90}Sr β source.

Experiment (b): G–M tube, scaler with timer, absorber set, γ-ray source.

INTRODUCTION:

A number of devices exist that can be used to detect radioactive particles. The Geiger counter is one of the easiest to use. It consists of a metal cylinder filled with an inert gas such as argon to a pressure of a few centimeters of mercury, together with additions of

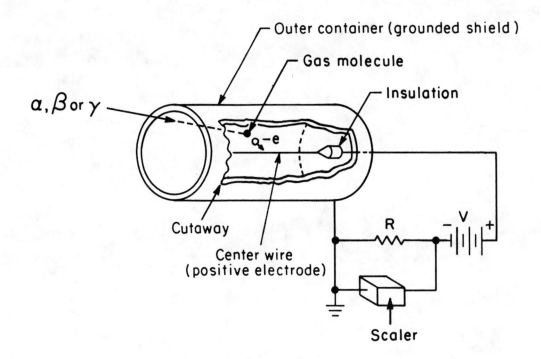

Fig.1. The Geiger counter.

halogens or organic vapors. The α or β particles, or γ rays, enter the cylinder through a thin window at one end. Gamma rays also can penetrate the metal walls directly.

A wire electrode runs through the center of the cylinder and is kept at a high positive potential (approximately 400 volts) relative to the conducting wall of the outer cylinder. When a high energy particle (α or β) or photon (γ) enters the cylinder it collides with and ionizes one or more gas molecules. The electrons produced from the gas molecules accelerate toward the central positive electrode, causing further ionization and more electrons. The ionization cascade continues until a large number of electrons are produced. These surge to the wire creating a current pulse through the resistor R. This pulse can be counted or made to produce a "click" in a loudspeaker. The number of counts (or clicks) is related to the number of nuclear disintegrations that initially produced the α or β particles, or γ rays.

The Geiger counter was first used by Hans Geiger, its inventor, in 1908 when he worked with Rutherford. Later modifications that increased the sensitivity of the device resulted in the instrument we now use, the Geiger-Mueller (G–M) counter or tube.

The Scaler. This is nothing more than an "electronic adding machine". It records the counts made by the (G–M) tube.

Rules for laboratory work with radioactive sources. Some of the following laboratory rules have been adapted from those in use by the Special Training Division, Oak Ridge Associated Universities.

1. Eating, drinking and the use of cosmetics in the laboratory are not permitted.

2. Keep all radioactive samples in their plastic container when not in use.

3. Keep the plastic container as far from you on the lab table as possible. (These sources are reasonably safe but just don't keep them "under your nose" when not in use.

4. As an added precaution use forceps when handling the γ ray source.

Experiment (a): Determining the operating voltage of a Geiger-Mueller counter.

For very low values of the applied voltage V in fig. 1, only very small ionization pulses are produced. But as the applied voltage is increased the electrons released in an ionizing event eventually acquire sufficient energy to ionize neutral molecules of the gas in the tube. As stated above, these electrons may cause further ionization and so on. This process is known as **gas amplification** and implies that the count rate of radioactive events builds up with the applied voltage. The region AB of the curve in fig. 2 illustrates this. Further increases in the applied voltage increase the gas amplification, the ionizing cascade of electrons spreading along the whole length of the central wire in the tube. Strong output pulses now result, although the count rate barely increases, no longer being proportional to

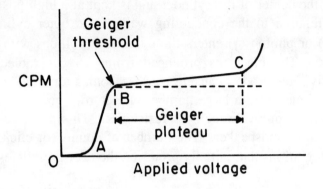

Fig.2. Count rate vs. applied voltage.

the applied voltage. This more or less horizontal region BC in fig. 2 is called the **Geiger plateau** and is the range used in (G–M) tube counting. Increasing the applied voltage beyond C initiates a continuous discharge which may seriously harm the tube.

All (G–M) tubes, even of the same size and type, do not operate at the same voltage because of differences in construction. It is necessary, therefore, to determine the correct operating voltage for the (G–M) tube you use. This is accomplished by determining the plateau region from a graph similar to fig. 2.

Experiment (b): Determination of the half value thickness of a gamma ray absorber.

Gamma rays are high energy electromagnetic radiations that appear during nuclear reactions. They have the same properties as X-ray photons of the same energy and travel at 3×10^8 m/s in a vacuum. One significant difference between X-rays and gamma rays is their source. X-rays are produced when electrons undergo deceleration or when orbital electrons jump from one orbit to another of lesser energy. On the other hand, γ-rays are produced by transitions between states of different energy within the nucleus. Although γ-rays react with matter in a variety of ways, in this experiment we are concerned primarily with their absorption by matter.

If a beam of gamma ray photons impinges on a sheet of absorbing material some of the radiation will pass on through while some will be absorbed or scattered. As the thickness of the absorber is increased, the fraction of the radiation passing through will decrease. For example, suppose N photons pass through a thickness x of absorber. The number of photons dN absorbed in an additional thickness dx is proportional to N and to dx. That is

$-dN \propto Ndx$.

The minus sign is used because an amount is <u>removed</u> from the γ-ray beam. Thus

$$dN = -\mu_x N dx \; , \tag{1}$$

μ_x being the linear absorption coefficient which is constant if γ-rays of only one energy are involved in the absorption. On examination of eq.(1) it is obvious that μ_x has dimensions of reciprocal distance.

A related constant which we will call μ, the mass absorption coefficient, is defined by the equation

$$\mu = \mu_x/\rho \tag{2}$$

where ρ is the density of the absorber. The quantity μ is usually expressed in units of cm^2/g. Then eq.(1) becomes

$$dN = -\rho\mu \, N dx \; . \tag{3}$$

The inverse of the mass absorption coefficient (μ^{-1}) is called the area density, and is expressed in g/cm^2.

Eqs.(3) or (1) may be rearranged to give greater clarity. If both sides of the equation are divided by N we obtain[1]

$$dN/N = -\rho\mu \, dx \; . \tag{4}$$

This equation expresses the fact that the fraction of the γ-rays that disappears in proportional to the corresponding increase in the thickness of the absorber. Eq.(4) may be integrated to give

$$N = N_0 e^{-\mu\rho x} \tag{5}$$

where N_0 is the number of photons entering the sample before any absorption has taken place. The exponent in eq.(5) is dimensionless, thus the product ρx is also the areal density, in units g/cm^2, of a thickness x of absorber. It is often used to describe the effective thickness. We can take the logarithm of both sides of eq.(5). This gives

---

[1]Equations of this form describe several processes in physics, life science, banking, sociology, etc., some with and some without the negative sign. With the negative sign dN/N represents a decrease while dx is an increase. When there is no negative sign dN/N increases with x. Of course, in these examples the variables N and x may represent entirely different things with different units. The positive sign often operates in population growth, where N may be number of people and x, time. In some aspects of banking, N = money and x = time. The amount of fossil fuel in the world and radioactive decay as a function of time are examples in which eq.(4) with the negative sign, is applicable.

$$\ln N = \ln N_0 - \mu\rho x \quad , \tag{6}$$

the equation of a straight line.

The half–value thickness or half–value length X. This is the value of x that will absorb 50% of the radiation. Eq.(6) is equivalent to

$$\ln(N/N_0) = -\rho\mu x \tag{7}$$

which simplifies to

$$X = \frac{\ln(2)}{\mu\rho} \quad . \tag{8}$$

In this experiment you will determine $\mu\rho$ by graphical analysis with eq.(6), and then calculate X, the half–value thickness, according to eq.(8).

PROCEDURE:

Experiment (a)

First we use a built-in test signal to verify the proper operation of the scaler.

1. The high voltage switches should be in the zero position and the power switch "OFF" before the line cord is plugged in.

2. Plug in the line cord and turn the POWER switch on. The 6 decade readout on the scaler will indicate a random count. Some digits may not even light up. To zero the readout, depress STOP; then depress RESET. The scaler will now show 000000.

3. Move the COUNT INTERVAL timer switch to the "one minute" position. Depress TEST switch. It will remain in the "ON" position.

4. Depress STOP and RESET. Both switches will return to the normal position but the red indicator lights will remain on. The 6 decade readout will indicate all zeros.

5. Increase the voltage by about 40V. Reset the counter and depress COUNT. Record the count after one minute.

6. Repeat step (5) until you reach about 600V. When the voltage reaches approximately 400V you will be in the vicinity of the threshold value of the scaler.

After you have finished using the scaler turn the high voltage adjusts to zero but do not turn off the scaler. If the scaler shows a 6 zero readout turn it off and on quickly so that a random number appears on the scale. This prolongs the life of the light emitting diodes (LED) that illuminate the zeros.

Experiment (b)

1. Set the scaler at the proper operating voltage. Remove all source from the counting area and take two or three 5-minute counts of the background radiation. This radiation comes from cosmic rays (which is primarily radiation from the sun), from radioactive atoms in our bodies and in the soil, and from other material around us such as bricks and metal used in building construction. This background radiation must be determined and subtracted from the sample activity.

2. Set the timer to count for one minute. Place the gamma ray source in the sample holder on the third shelf (from the top) of the tube stand.

3. Place the empty absorber slide in the shelf immediately above the sample.

4. Determine the activity of the sample with no absorber, and record this value in the data table.

5. Insert the thinnest polythelene absorber slide in the second shelf and determine the activity. Record the absorber thickness as indicated in the absorber set.

6. Repeat step (5) for all of the polythelene slides. If time permits, repeat the experiment for the four lead absorbers.

<u>ANALYSIS</u>:

Experiment (a)
1. Plot a graph of Counts per Minute (CPM) vs. Voltage. Your graph should resemble fig. 2.

2. Record the I.D. number of the scaler, and estimate the threshold value (It should be within the first quarter of the plateau region). Record this value in the data table (table 1) as the operating voltage of the scaler.

Experiment (b)
1. Plot a graph of ln (Corrected Activity) vs. Polythelene Absorber Thickness on linear paper, and draw the best straight line through the data.

2. Determine the slope $\mu\rho$. Use eq.(8) and your value for the slope to determine the half-value thickness X.

Apparatus No:_____

Operating voltage:_____V

Counting Time:_____minutes

Background Radiation:_____CPM

Absorber Material:_____(Polythelene or lead)

Absorber No.	Absorber Thickness g/cm²	Total Count	CPM	Corrected CPM

Table 1. Gamma ray absorption

3. Using the graph, find the absorber thickness at a corrected CPM value of $N_0/2$. Compare this value for the half–value thickness to the one determined from the slope.

4. Record the results of all the groups in your lab. Find the average value and calculate the deviation of your value from the mean.

5. Repeat steps (1) through (4) for the lead absorber.

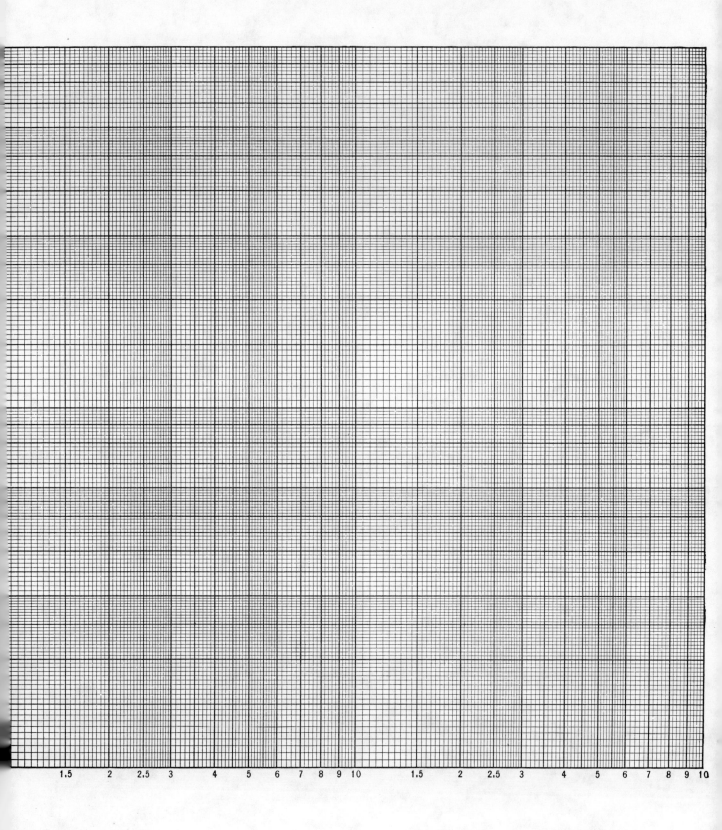

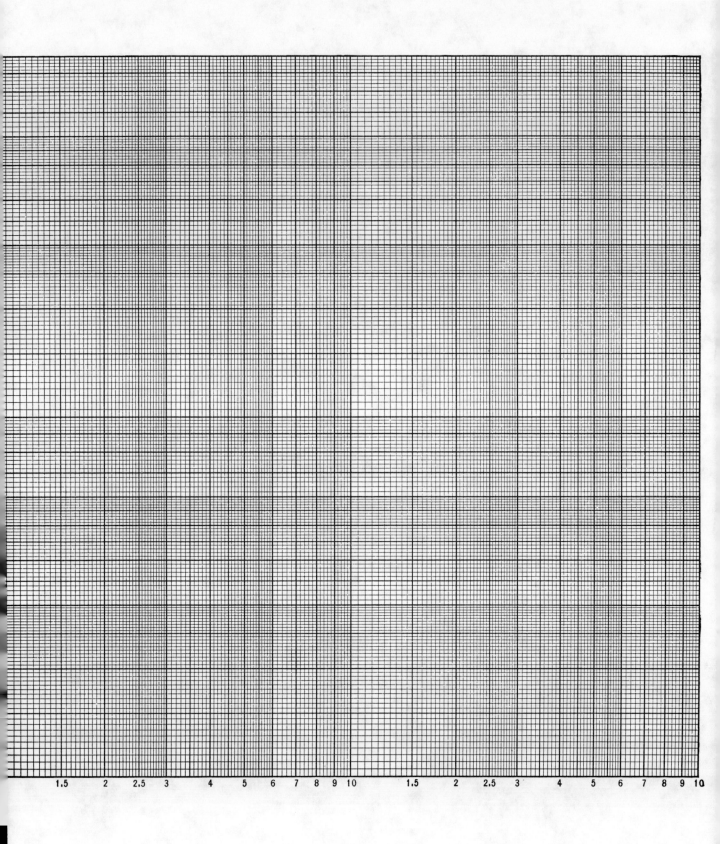

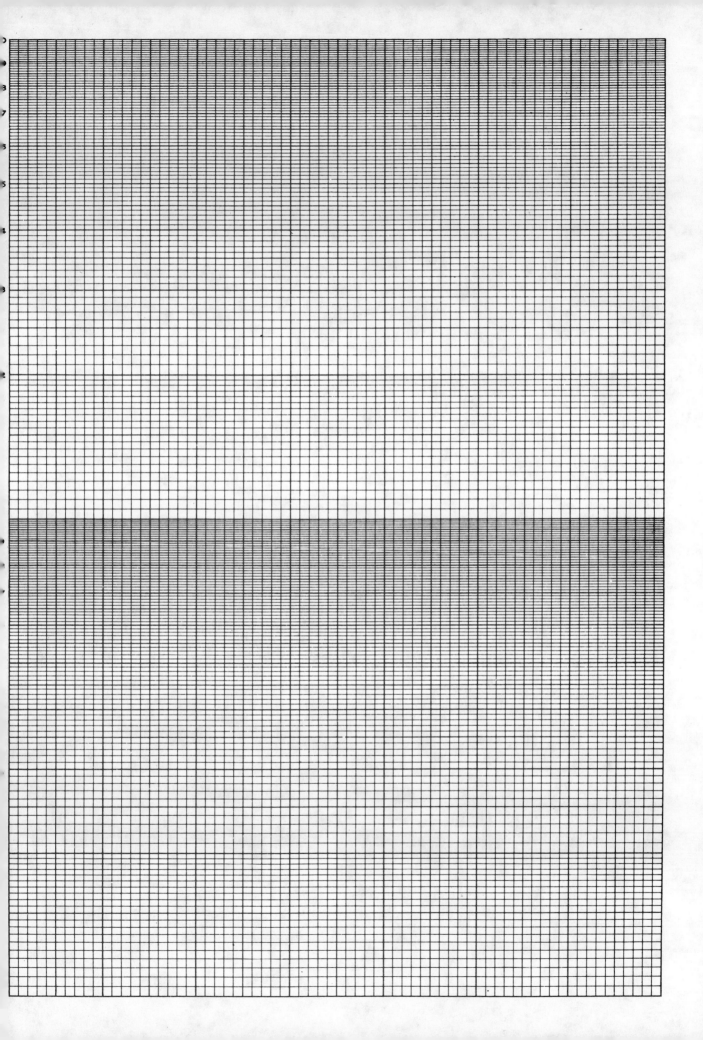

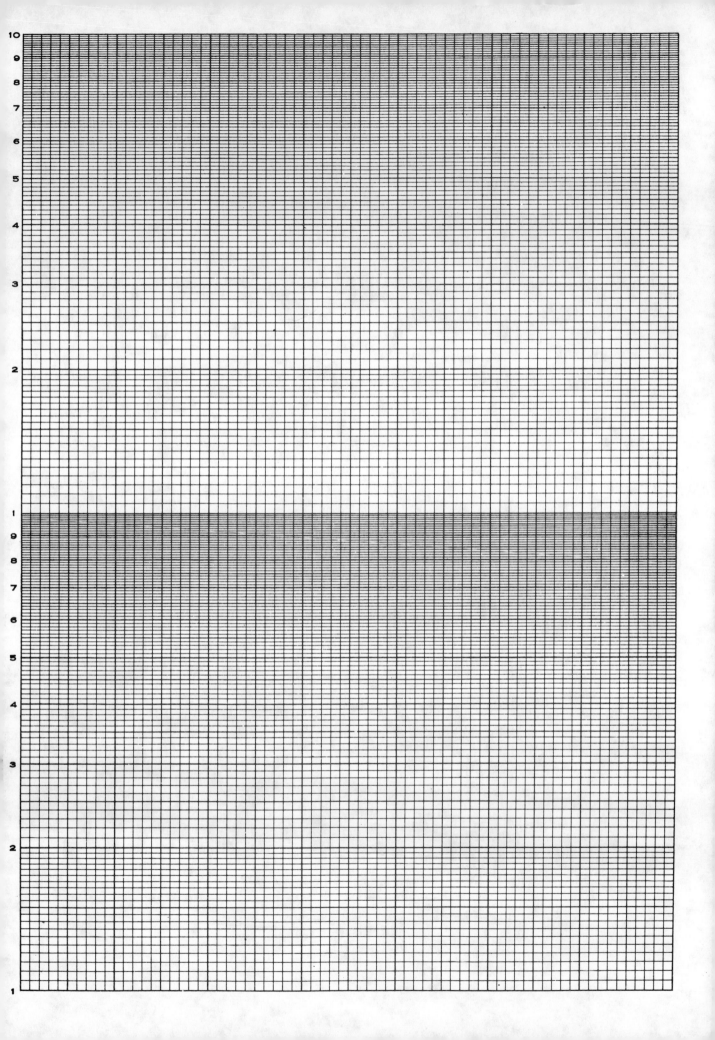

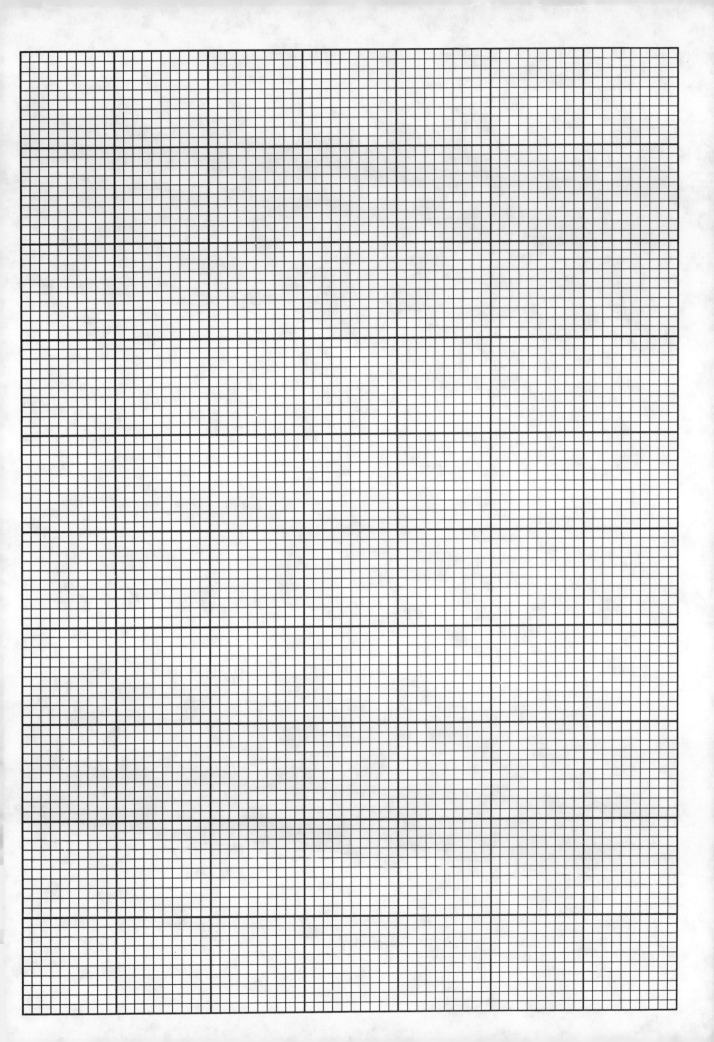

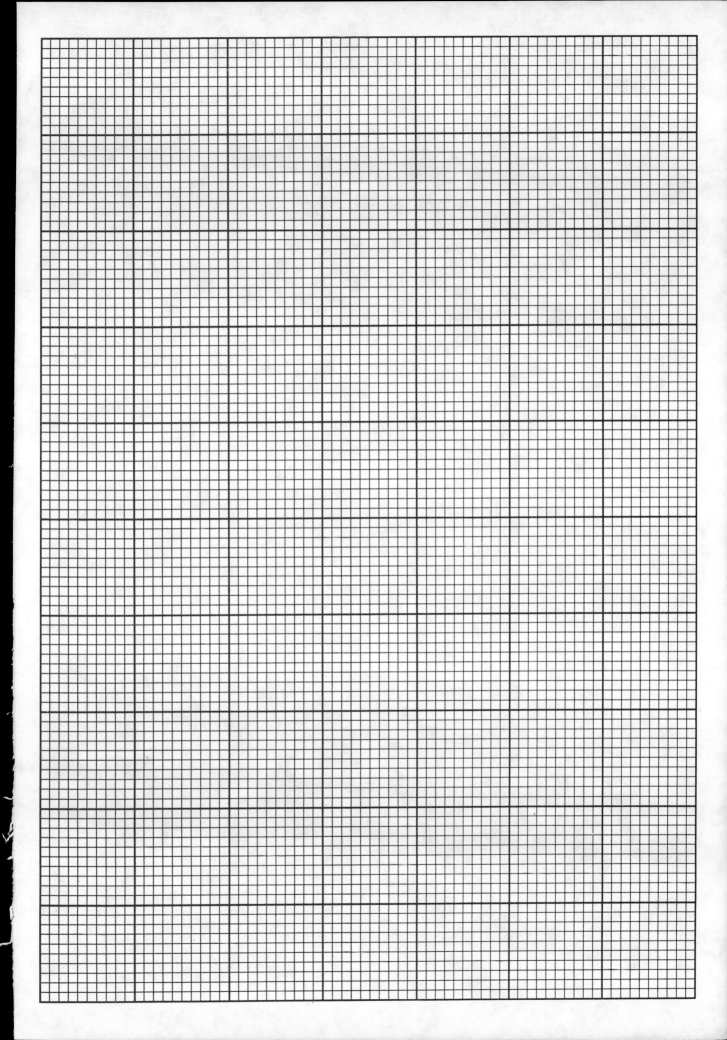

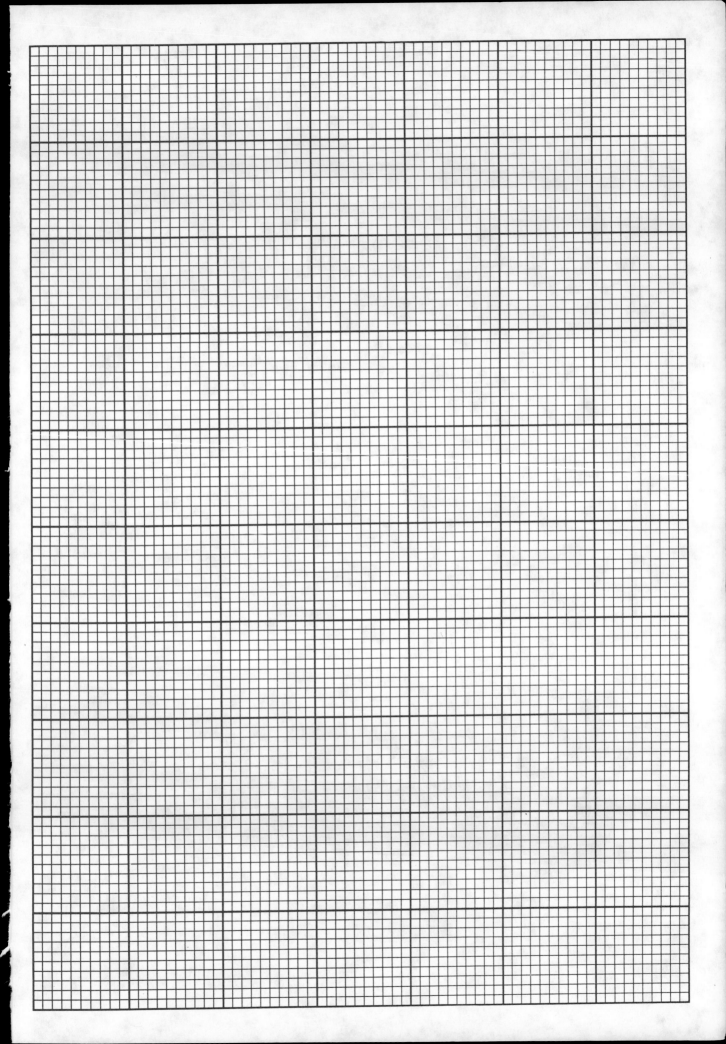

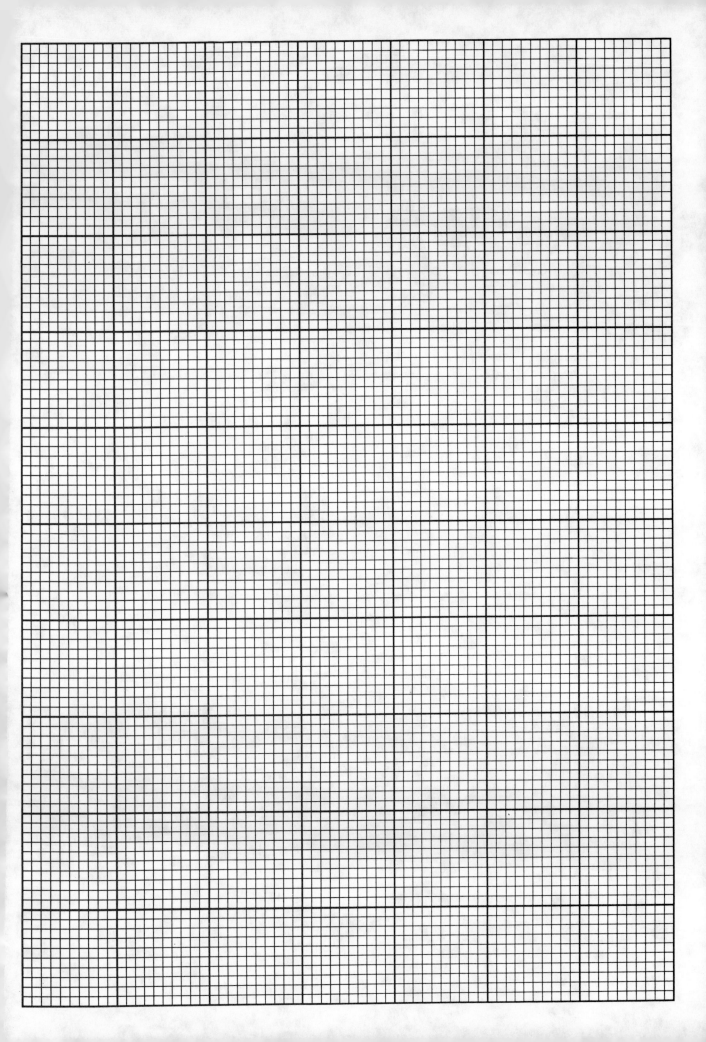

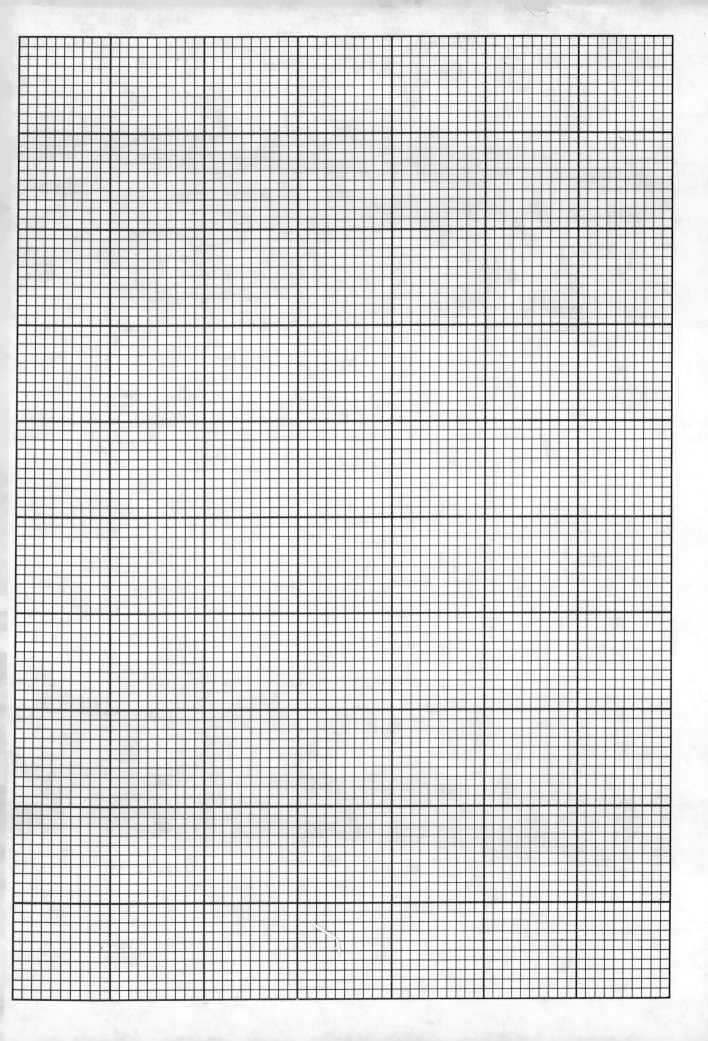